Lecture Notes in Bioinformatics

Subseries of Lecture Notes in Computer Science

Jasmin Fisher (Ed.)

Formal Methods in Systems Biology

First International Workshop, FMSB 2008
Cambridge, UK, June 4-5, 2008
Proceedings

 Springer

Series Editors

Sorin Istrail, Brown University, Providence, RI, USA
Pavel Pevzner, University of California, San Diego, CA, USA
Michael Waterman, University of Southern California, Los Angeles, CA, USA

Volume Editor

Jasmin Fisher
Microsoft Research
Computational Biology Division
7 J J Thomson Ave., CB3 0FB, Cambridge, UK
E-mail: Jasmin.Fisher@microsoft.com

Library of Congress Control Number: 2008927203

CR Subject Classification (1998): I.6, D.2.4, J.3, H.2.8, F.1.1

LNCS Sublibrary: SL 8 – Bioinformatics

ISSN   0302-9743
ISBN-10   3-540-68410-7 Springer Berlin Heidelberg New York
ISBN-13   978-3-540-68410-7 Springer Berlin Heidelberg New York

Springer is a part of Springer Science+Business Media

springer.com

© Springer-Verlag Berlin Heidelberg 2008

Typesetting: Camera-ready by author, data conversion by Scientific Publishing Services, Chennai, India
Printed on acid-free paper   SPIN: 12273294   06/3180   5 4 3 2 1 0

Preface

This volume contains the proceedings of the first international meeting on Formal Methods in Systems Biology, held at Microsoft Research, Cambridge, UK, June 4–5, 2008.

While there are several venues that cover computational methods in systems biology, there is to date no single conference that brings together the application of the range of formal methods in biology. Therefore, convening such a meeting could prove extremely productive. The purpose of this meeting was to identify techniques for the specification, development and verification of biological models. It also focused on the design of tools to execute and analyze biological models in ways that can significantly advance our understanding of biological systems. As a forum for this discussion we invited key scientists in the area of formal methods to this unique meeting.

Although this was a one-off meeting, we are exploring the possibility of this forming the first of what might become an annual conference. Presentations at the meeting were by invitation only; future meetings are expected to operate on a submission and review basis.

The Steering Committee and additional referees reviewed the invited papers. Each submission was evaluated by at least two referees. The volume includes nine invited contributions.

Formal Methods in Systems Biology 2008 was made possible by the contribution and dedication of many people. First of all, we would like to thank all the authors who submitted papers. Secondly, we would like to thank our additional invited speakers and participants. We would also like to thank the members of the Steering Committee for their valuable comments. Finally, we acknowledge the help of the administrative and technical staff at the Microsoft Research Cambridge lab.

April 2008   Jasmin Fisher

Organization

Steering Committee

Luca Cardelli (Microsoft Research, Cambridge, UK)
David Harel (Weizmann Institute of Science, Israel)
Thomas A. Henzinger (Ecole Polytechnique Fédéral de Lausanne, Switzerland)
Jasmin Fisher (Microsoft Research, Cambridge, UK)
Amir Pnueli (New York University, NY, USA)
Claire Tomlin (Stanford, CA, USA and UC Berkeley, CA, USA)
Stephen Emmott (Microsoft Research, Cambridge, UK)

Referees

L. Cardelli	J. Fisher	T.A. Henzinger
S. Emmott	D. Harel	N. Piterman

Organization

Conference Chair	Jasmin Fisher (Microsoft Research, Cambridge, UK)
Local Organization	Hayley North (Microsoft Research, Cambridge, UK)
Web Page Design	Nick Duffield (Microsoft Research, Cambridge, UK)

Sponsoring Institution

Microsoft Research, Cambridge, UK

Table of Contents

Contributed Papers

Generic Reactive Animation:
Realistic Modeling of Complex Natural Systems

David Harel and Yaki Setty

Department of Computer Science and Applied Mathematics,
The Weizmann Institute of Science,
Rehovot 76100, Israel
{david.harel,yaki.setty}@weizmann.ac.il

Abstract. Natural systems, such as organs and organisms, are large-scale complex systems with numerous elements and interactions. Modeling such systems can lead to better understanding thereof and may help in efforts to save on resources and development time. In recent years, our group has been involved in modeling and understanding biological systems, which are perhaps the prime example of highly complex and reactive large-scale systems. To handle their complexity, we developed a technique called reactive animation (RA), which smoothly connects a reactive system engine to an animation tool, and which has been described in earlier publications. In the present paper we show how the basic idea of RA can be made generic, by providing a simple general way to link up any number of reactive system engines — even ones that are quite different in nature — to an animation tool. This results in natural-looking, fully interactive 3D animations, driven by complex reactive systems running in the background. We illustrate this with two examples that link several tools: Rhapsody for state-based specification, the Play-Engine for scenario-based specification, MATLAB for mathematical analysis and the 3DGAMESTUDIO for animation. Our examples are both from biology (pancreatic development) and from everyday activities (e.g., gym training).

1 Introduction

Natural systems, such as organs and organisms, are large-scale complex systems that maintain an ongoing interaction with their environment and can thus be beneficially specified as reactive systems [18,23]. This observation has led to quite a lot of work on modeling biology using various software engineering tools and ideas to simulate behaviors in various natural systems. To handle the complexity of natural systems we need even better visualization techniques than those

---

[1] This research was supported in part by The John von Neumann Minerva Center for the Development of Reactive Systems, and by a grant from the Kahn Fund for Systems Biology, both at the Weizmann Institute of Science. Part of this David Harel's work carried out during a visit to the School of Informatics at the University of Edinburgh, which was supported by a grant from the EPSRC.

used for conventional reactive systems. Indeed, animating natural system reveals unexpected behavioral properties that emerge from the simulations at run-time. These *emergent properties* are not explicitly programmed, but are often a consequence of massively concurrent execution of basic elements that act in concert as a population [8].

In recent years, our group has been involved in developing a technique called *reactive animation*, whereby reactive systems whose external form requires more than conventional kinds of GUIs are modeled by languages and tools for the reactivity linked with tools for true animation. In our earlier work on this topic [10] we were motivated by a complex modeling example from biology, which required us to model many thousands of T cells developing, moving around and interacting in the thymus gland, and we wanted the animation to show this actually happening. Our implementation linked the Rhapsody tool, with its statechart model of the system, to a Flash animation front-end, and enabled interaction of the user with the system either via Rhapsody or via the Flash front-end. In that work, the animation was two-dimensional. More importantly, however, the linking itself was binary — one reactive system engine to one animation tool — and it was carried out in an ad hoc fashion. It thus did not provide a mechanism for a more generic reactive animation.

In this paper we exhibit a stronger kind of reactive animation, by devising a specific mechanism for linking any number of tools, which may include an animation tool and reactive system engines of different kinds. We illustrate the technique and the underlying principles (as well as the feasibility of 3D reactive animation) by linking up the Rhapsody tool, which supports statecharts and object diagrams, and the Play-Engine [22], which supports live sequence charts (LSCs)[9], to 3DGameStudio (3DGS) [1] for animation and MATLAB[28] for mathematical analysis. Our two reactive engines, Rhapsody and the Play-Engine, which follow state-based and scenario-based approaches, respectively, are connected through a central mechanism with the 3DGS animation tool and the Mathematical GUI. The mechanism is general enough to support and maintain any number of such links. We demonstrate the architecture using two main examples, a complex one from biology and a simpler one closer to everyday life. We will also briefly discuss possible future directions for modeling other natural systems.

This paper is supplemented by a website (http://www.wisdom.weizmann.ac.il/~yaki/GRA), which contains a recording of a short session that was carried out with the different tools linked together. The material on the site also includes several more detailed video clips showing some of the possibilities of the setup. We have deliberately left out in this paper much of the technicalities of the method and its implementation, and have focused instead on the motivation and objectives, and on examples. For readers who are interested in the technical details, there is a technical report that explains things in greater detail, and which can be downloaded from the above website. That report also includes a detailed description of how the architecture is employed, and it briefly suggests directions for standardizing the platform. The site also points to some additional supplementary material on the examples (e.g., explanatory clips and interactive illustrations).

In our first example, we employ the architecture to model pancreatic development in the mouse, which besides its biological importance turns out to be a highly complex system for modeling, with numerous different kinds of objects. Our model includes a state-based specification linked to a 3-dimensional animated front-end and a mathematical analysis GUI [36]. A prerecorded clip of the simulation at run time is available at www.wisdom.weizmann.ac.il/~yaki/runs. In this example, Statecharts [16] (in Rhapsody [38]) describe the behavior of the biological objects themselves (e.g., cells), which are represented in the front-end as 3D elements possessing realistic animation attributes. Statistics and analysis of the simulation are shown in a separate GUI (in MATLAB). We discuss how available scientific data is specified as statecharts in Rhapsody, and how scenarios via the LSCs and the Play-Engine may answer the need for mutating the system. The front-end shows the animation continuously, and provides the means to interact with it. The mathematical analysis GUI provides statistics and graphs of the simulation. Generally, the simulation corresponded well with the biology, indicating that the 3D structure emerging from the simulation seems to capture pancreatic morphogenesis in mice. Moreover, this platform enabled to perform a set of *in silico* experiments, which reproduced results similar to in-vivo efforts and provided a dynamic description. In addition, the model suggested new intriguing results that are currently being tested through collaboration, for an experimentally validation.

Our second illustration of the method is a more intuitive running example (pun intended...) of a 3-participant gym training system, which includes a team leader and two team members, running, walking, jumping, crawling and standing, and if needed also swimming and wading (in the special case of scenarios involving flooding). The system also includes a moving camera, sub-viewing abilities, and more. We discuss the way certain parts of the overall controlling behavior are specified in scenarios via the LSCs and the Play-Engine, whereas others, such as the behavior of the participants themselves, are specified using statecharts in Rhapsody. The front-end shows the animation continuously, and provides the means to interact with it. Interactive illustrations of this example are available at http://www.wisdom.weizmann.ac.il/~yaki/GRA/gym.

2   Reactive Animation

Reactive animation (RA) [10] is a technology aimed at combining state-of-the art reactivity with state-of-the-art animation (Fig. 1A). RA links the effort of reactive system design and the front-end design by bridging the power of tools in the two separate areas. In essence, RA has two arms: One comprises powerful tools and methods for reactive systems development, the heart of which is a rigorous specification of the systems reactivity. The other comprises powerful animation tools to represent the specification as an intuitive, controllable, animated front-end. The animated front-end serves as a communication channel for better human understanding of the simulation. Technically, RA is based on the view that says that a system is a closely linked combination of what it does and what it looks like. From this stem two separate but closely linked paths: reactive behavior design and front-end design.

Implementing a tool for RA poses a number of challenges: accuracy, performance (e.g., CPU usage, memory management, smoothness of the resulting animations), distribution, ease of interaction, openness and platform independence. These should be considered at the architecture level and with the specific functionality chosen.

RA was conceived of during an effort to model and simulate the development of T-cells in the Thymus gland [10], where it was implemented using a direct communication socket between a state-based model (using statecharts in Rhapsody) and a 2D animated front-end (using Flash). Our work improves upon [10] by providing a generic, modular and fully distributed multi-party architecture for RA, which also employs 3D animation (Fig. 1B).

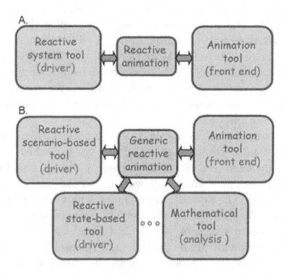

Fig. 1. A. Reactive animation: an animation tool is binary linked to a reactive system tool to enable natural-looking, fully interactive animations. B. Implementing generic multi-party reactive animation: reactive engines, animation, mathematical and any other type of tool are linked together using a central routing server (a star topology). Tools communicate through a TCP socket and transmit XML based messages, enabling a fully distributed, platform-independent implementation (in a way similar to Soap).

3   Implementation Architecture

The architecture of our examples includes two reactive engines supporting two different modeling approaches: a state-based, intra-object approach, and a scenario-based, inter-object approach. The reactive engines are linked up to a three-dimensional animated front-end and a mathematical analysis GUI, using a central routing server. In principle, the architecture can be extended by using any number of additional components of any related kind (e.g., a 2D front-end using Flash).[1] Below we briefly describe each of the architectural components, a detailed technical report is available at www.wisdom.weizmann.ac.il/~yaki/GRA/.

The Central Routing Server: The server was implemented as a multi-threaded executable application. Each thread serves as a communication plug-in for one architectural component. A TCP socket is initialized upon registration to the server, enabling message transmission and XML parsing. See the supplementary technical report for a more detailed description.

The State-Based Specification: We use the language of statecharts [17] and the Rhapsody tool [38] to implement state-based specification. Statecharts are naturally suited for the specification of objects that have clear internal behavior, an attribute we call *intra-object*. Together with object model diagrams, they provide a graphical representation of the dynamics of objects using states, transitions, events, and conditions [20]. The language makes it possible to visualize the behavior of an object in a way that emphasizes the elements in its life-cycle. Rhapsody is a model-driven development environment supporting statecharts and object model diagrams (see [20]), and can be viewed also as a UML tool. It enables object-oriented design, with full execution of the statechart-rich models, and full code generation. Table 2, top details a representative example for a state-based specification.

The Scenario-Based Specification: We use the language of LSCs, live sequence charts [9] and the Play-Engine [22] to implement scenario-based specification. LSCs are scenario-based, and inter-object in nature, and are particularly fitting for describing behavioral requirements. LSCs extend classical message sequence charts (MSCs) with logical modalities, depicted as hot and cold elements in the charts. The language thus achieves far greater expressive power than MSCs, and is comparable to that of temporal logic. In particular, LSCs can specify possible, mandatory and forbidden scenarios, and can be viewed as specifying multi-modal restrictions over all possible system runs. An LSC typically contains a prechart and a main chart. The semantics is that if the scenario in the prechart executes successfully, then the system is to satisfy the scenario given in the main chart. The Play-Engine is the tool built to support LSCs, that enables a system designer to capture behavioral requirements by 'playing in' the behavior of the target system, and to execute the specified behavior by 'playing out'. In the play-out phase the user or an external component executes the application as if it were the real system. Table 2, bottom details a representative example of a scenario-based specification.

The 3D Animated Front-End: The front-end was implemented using a three dimensional authoring tool (3D Game Studio (3DGS) [1]), which supports real time rendering of 3D animation. In 3DGS, objects can have associated actions, which appear as part of its attributes. The scripting language of 3DGS, C-Script,

---

[1] In earlier work of our group, we developed InterPlay, which is a different kind of technique to connect reactive system engines, based on pairings of connections [4].

enables control of animation objects (e.g., changing an attribute) and supports object oriented programming. We choose to add a controlling GUI to our system as part of the front-end (as was done also in the thymus model [10]). However, in more complex systems such a GUI could very well complicate the front-end and should be designed as a different architectural component, possibly using a GUI-building tool.

Mathematical Analysis GUI: We designed a mathematical analysis GUI using MATLAB from MathWorks [28], which is a high-level language and interactive environment for computational tasks. This GUI generates various graphs and statistics based on data received from the simulation. Similar mathematical analysis tools such as MATHEMATICA from Wolfram Research, can also be plugged-in the architecture.

The Architecture at Run-Time: Each component, when executed, initiates a connection with the Central Routing Server. The setting of components in the architecture enables pairwise message transfer between them. At run-time, message passing drives the simulation in the participating components. For example, messages from the reactive engines (i.e., Rhapsody and Play-Engine) drive the animation in the front-end. Table 1 describes in detail possible runs of one of our examples.

4   Modeling a Large-Scale Biological System

We have employed the proposed generic RA setup to model the development of the pancreas, a highly complex system, containing numerous objects. The pancreas is an essential organ, which is involved in regulation of metabolic and digestive pathways. During development, it takes on an interesting three-dimensional cauliflower-like shape. A prerecorded run of the simulation is available at www.wisdom.weizmann.ac.il/~yaki/runs

Abnormal functioning of the pancreas leads to lethal diseases such as pancreatitis and diabetes. Our model includes a comprehensive state-based specification that results from analyzed scientific data. It is linked to an animated front-end and a mathematical analysis GUI. In the future, we plan to link these two also with a scenario-based specification to simulate mutations (e.g., defective blood vessel). In [36], we provide details of this the work and discuss the intriguing novel ideas that emerged from the model. See http://www.wisdom.weizmann.ac.il/~yaki/abstract/ for a short description.

Modeling Pancreatic Development: We modeled pancreatic development as an autonomous agent system [5] in which cells are autonomous entities that sense the environment and act accordingly. The cell object in the model consists of three elements, the nucleus, the membrane and the cell itself. The nucleus operates as an internal signaling unit that expresses genes to drive the development, while the membrane acts as an external signaling unit that senses the environment and alerts the cell. The cell itself changes states in response to the various signals (see Fig. 2). The environment was modeled as a computational

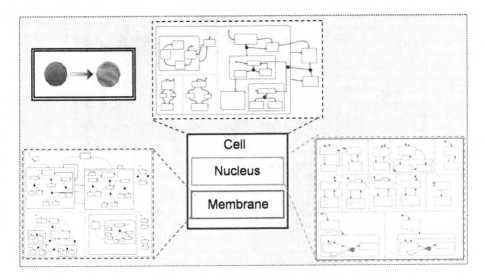

Fig. 2. The model for a cell as an autonomous agent accompanied with its visualization (top-left)

grid, with values that designate concentrations of biological factors. Various biological components participate in the process by regulating factors in the environment. Each of these was specified as an object accompanied by a statechart to describe its behavior. Cells however, are considered as the basic objects, and the progress of the simulation/execution relies very much on their behavior. An execution of the model is initiated with approximately 500 cells, which, among other processes, proliferate and create new instances. A typical execution ends with almost 10, 000 objects.

Designing an Animated Front-End: We visualize the simulation in an animated front-end, which we built based on what is depicted in the literature. Each one of the participating components is represented as a 3D element possessing attributes to represent change in location and behavior (Fig. 3A). For example, the cells are represented in the front-end as spheres. At run time, an instance of a cell directs its corresponding animated sphere according to its active state. A differentiated cell might, for example, change its color to depict the new stage. As the simulation advances, the cells dynamically act in concert to form the cauliflower-shape structure of the pancreas (Fig. 3B). At any stage, the user can halt the simulation and query objects or interact directly with the emerging structure (e.g., 'slice' the structure) (Fig. 3C).

Mathematical Analysis GUI: We designed a GUI in MATLAB to provide mathematical analysis of the simulation. The GUI continuously receives data from the reactive engine, analyzes it and provides graphs and statistics (for example, graphs of cell population, proliferation rate etc.). It thus enables the user to evaluate the dynamics of the simulation over time. Fig. 3D shows a snapshot of a graph that displays the cell count over time. In general, the system

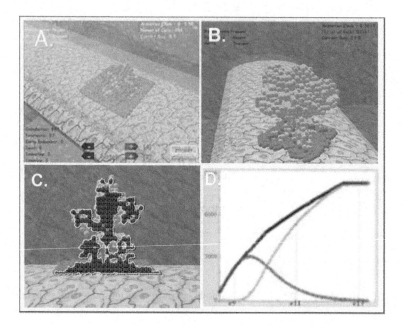

Fig. 3. A. 3D Animated front-end for the pancreatic development. B. The pancreatic structure emerging from the simulation. C. User interaction with the simulation, the simulation was halted and a cross-section cutting was triggered. D. Mathematical analysis of the pancreatic development: number of cells as function of time.

developer can design many graphs and statistics, which are related to the relevant system and whose data is gleaned from the model.

Specifying Mutations in the Simulation: In the pancreas project, as well as in modeling other large-scale natural systems, scenario-based programming may help in specifying mutations in a rather natural way. A typical mutation scenario should specify the changes in the model that correspond to relevant mutations between elements in the system. At run-time, the user may trigger relevant scenarios to mutate the system and then watch the effect on the simulation. For example, a mutation scenario in pancreatic organogenesis would specify the effect of a defective blood vessel on the participating elements, in particular cells.

In the past, work in our group has shown scenario-based programming to be beneficial for modeling biology. For example, it was the dominant tool used for the specification and verification of certain developmental aspects in the reproductive system of the *C. elegans* nematode (see e.g., [26,27,35,25]).

The Pancreas at Run-Time: Once the model is executed, instances of the Cell are created and appear in the front-end as a sheet of red spheres on the proper location on the flat endodermal Gut. Once a Cell instance is created, one state in each concurrent component of its statechart is set to be in active state. At this point, the Cells are uniform and their active states are set to the initial states . In parallel, the environment is initiated and defines the initial concentrations of factors in the extracellular space. As the simulation advances, cells respond to various events (e.g., the concentration of factors in their close vicinity) by changing their active states accordingly. Hence, the sheet loses uniformity at a very early stage of the simulation.

As the simulation advances, among other things, cells are differentiated, proliferated and move. The processes are driven by many extracellular events (e.g., from the membrane) and intra-cellular events (i.e., from the nucleus). These events change the active states in orthogonal specifications through the various stages of the cell's life cycle. For example, the proliferation process is initiated by extra-cellular signals when the membrane senses relevant factors and generates a chain of inter-cellular events (in the cell and the nucleus) that promote cell division. Proliferation ends when the Cell duplicates itself by creating an identical instance. In turn, a message is sent to the front end, which creates a new identical sphere corresponding to the new Cell at the proper location.

The cell population acts in concert to drive the simulation, by promoting various decisions in individual cells. Consequently, messages are continuously being sent between the different components in the architecture, in order to drive the simulation. The process is displayed in parallel in each one of the components. The state-based specification highlights the active state of the different objects and, at the same time, the front-end and the mathematical GUI continuously visualize and analyze the simulation.

Once scenario-based specification is combined to the model, interplay between scenario- and state-based specifications will drive mutations in the simulation. The user would be able to trigger mutations through the various tools in the model and then watch the mutated behavior through the animation and analysis. Such mutations may be tested in vivo on the real system through laboratory collaboration.

5   A Team Training Model

The second example we describe here is a simple, yet representative, model. It involves gym training sessions for a team of three: a team leader and two team members. The team leader performs various actions at different speeds, and the team members follow suit, after a short "comprehension" delay. Team members, however, are not as physically fit as the team leader, and need to rest while performing certain fast actions. In addition, the team leader reacts to environmental changes (e.g., a Flood) and performs an appropriate set of actions to handle such situations. The model implements a state-based and scenario-based model, linked to an animated front-end. Detailed descriptions of two execution examples of the model are given in Table1. Also, http://www.wisdom.weizmann.ac.il/~yaki/ GRA/gym contains several self-explanatory video clips.

Modeling Team Behavior Using Statecharts: We used the state-based approach to specify the team's actual behavior. Our Rhapsody model includes three classes: the team leader, the team member and the team. We demonstrate statecharts using two team members, however, the model can be easily extended to handle any number thereof. The behavior of each model element is specified by a different statechart. The statecharts of the leader and a member describe the action and the speed taken and are very similar (see Table2 for greater details). The statechart of the team handles interactions with the environment.

Specifying Team Training Tasks Using LSCs: We used the scenario-based approach to specify training tasks for the team. The Play-Engine model specifies several training task, which include a set of instructions for the team and for the environment. Four different tasks have been specified: Escape, Flood, Storm and Volcano Eruption. Each of these initiates a different scenario. Triggering a new task while another is being executed causes a violation and the new LSC terminates. An instruction message (e.g., crawl, jump, swim) triggers an LSC that forces the engine to execute a specified set of messages. In addition, LSCs specify camera control and environmental changes (see Table2 for greater details).

Designing an Animated Front-End: The front-end for the model consists of a real-time 3D animation of the training exercise. Animated renditions of a team leader and a team member were created based on the cbabe model of 3DGS. The participants perform various actions such as walking, swimming, wading etc. The user may query the model by clicking an animated figure, and the relevant data (e.g., ID, current action) is displayed next to it. Environmental changes such as a flood may occur and the animated front-end displays the change accompanied by matching sound effects. The GUI portion of the front-end enables the user to trigger an instruction to the team (e.g., to jump fast) or to assign a task (e.g., to escape), respectively. He or she also controls the camera's activity and may query the running simulation.

Team Training at Run-Time: Once the model is executed, the state-based specification of the team behavior is initiated and the active state of each player is set to Stand (i.e., the initial state). As the simulation advances, various events (either internal or external) change the active state of the players. For example, instructions from the leader to the members are internal events of a team that drive within the state-based specification. Similarly, external events from the scenario-based specification, describing team tasks (e.g., flood), directs the active state of state-based specification.

The behavior of the model is continuously visualized at the front-end in various manners. For example, when a flood task is initiated, a water wave appears at the front-end and the team starts wading. At every time-point, the user may interact with the model so as to trigger different behaviors of the team (e.g., change the action of team members).

6   Additional RA Examples

We are currently looking into using our techniques in additional biological modeling projects. We use the generic reactive animation architecture to visualize behaviors as part of the GemCell project [2]. GemCell contains a generic statechart model of cell behavior, which captures the five main aspects of cell behavior (proliferation, death, movement, import and export). This generic model is coupled with a database of biological specifics (DBS), which holds the information about the specific cellular system. Modeling a particular segment of biology involves setting up the DBS to contain data about the specific behaviors and responses of the particular kinds of cells in the system under description. During execution, statecharts read in the specific data and the combination runs just as described in the model above.

We have employed the generic reactive animation architecture to link the GemCell model (in Rhapsody) with a 2D animated from-end (in Flash). At run time, the front-end continuously visualizes the behavior of numerous cells. The visualization clarifies the underlying principles in simulation. This project is still in its early stages of development.

In addition to this, we have tested the genericity of our architecture on some other examples, with different tools. We have a 'traffic handing' model that uses our architecture to link the scenario-based programming tool, the Play-Engine, to a 2D animated front-end using Flash. Also, we have designed a 'police at work' game using S2A[21], an aspectJ-based tool for scenario-based programming, linked to a 2D animated front-end in Flash.

Further details of these examples appear in http://www.wisdom.weizmann. ac.il/∼yaki/GRA.

7 Discussion

In the last few years, increasing interdisciplinary work combines experimental results with theoretical models in order to explain various natural systems (see e.g., [3,7,15,29]). Another type of modeling work formalizes gene expression and protein activity using a variety of mathematical and computational tools (for example, see [6,24,31,32,33]). However, most of the relevant work ignores multiple concurrent activities and focuses on a single mechanism in the entire system. An example for comprehensive modeling is the beating heart project [30], which formalizes the electric activities in the heart. However, by its mathematical nature, the model interactivity and real time animation are limited since simulations require much computation time.

Recently, various work uses computational modeling approaches for natural systems. In [14], hybrid automata are used to model the Delta-Notch mechanism, which directs differentiation in various natural systems. In [12], computational challenges of systems biology are described and various approaches for achieving them are discussed. A similar motivation for model-driven engineering approaches is discussed in [34]. Recently, in [13] computational and mathematical approaches are reviewed and the term *executable biology* is used to describe the kinds of modeling carried out in our group, and recently also elsewhere. In [39], a model for a eukaryotic cell is built, in which a UML class diagram was used to formalize the relations between a cell and its sub-cellular elements. The setup was empowered by specifying behavior of different cell types (e.g., red blood cell)

Table 1. Illustration of two execution examples of the team training model (for interactive illustration and recorded clips see `http://www.wisdom.weizmann.ac.il/~yaki/GRA/gym`)

Illustration of the Architecture:

The Team Training Model: Message transmission between the architectural components of the model: The state-based model (ST, at top left); The scenario-based specification (SC, at bottom left); and the animated front-end (FE, at middle right).

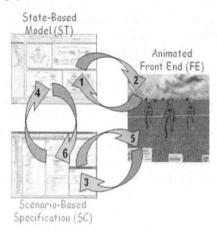

Example 1: Medium Speed Jumping	**Example 2:** Flood Training Task
Participating components: FE, ST	**Participating components:** FE, SC,ST
Description:	**Description:**
When the user sets the speed scroll bar to medium and clicks the `jump` button, FE notifies (i.e., sends a message to) ST about the instruction (1). Accordingly, the `team` object in ST generates an inner event to set the statechart of the team leader to jump at medium speed (i.e., the `Action` sub-statechart moves to the `Jumping` state and the `Speed` sub-statechart is set to `Medium`). After a predefined time interval, the `team` object generates another event for the statecharts of each of the team members. Upon entering the `Jumping` state, ST notifies FE to animate the action (2). Consequently, the animated team leader jumps, and the team members follow suit. The user can change the camera's position to view the team from different angles. Camera relocation, however, has no effect on the running simulation. During a run of the system, the user may query the model or relocate the camera. When a query is requested, FE notifies ST (1), which provides the appropriate information. Changes in the camera, however, do not interact with the reactive engines.	The user assigns a flood task to the team by clicking on the appropriate button. Consequently, FE notifies SC and the `flood` LSC is activated (3), initiating the scenario. The first message in the `flood` LSC is a running instruction for the team. Consequently, the `run` LSC is activated, and it notifies ST to run at a slow speed (4). Later on, a message in the `flood` LSC instructs the environment to initiate a flood. The `FloodAlert` LSC is activated and notifies FE (5). Accordingly, a water layer is displayed and a corresponding splashing sound is played. Immediately after this, the `swim` LSC is activated and it notifies ST. The `swim` message enables a forbidden element in the `run` LSC, causing the `run` LSC to exit. Again, the `swim` LSC notifies ST, which in turn notifies FE(2). The `flood` LSC completes after it instructs the team to walk, the environment to end the flood, and the team to stand (i.e., to stop moving). At this point, there are no more LSCs active in SC, the statecharts of ST are all in the `standing` state, and the animated figures in FE are standing, ready for the next task. In case a fast speed action is taken, each of the two `team member` objects will enter a `resting` state after some time. Consequently, ST notifies SC(6) and FE(2). Concurrently, the team member in FE changes its appearance to resting, an SC in SC is triggered, and it notifies FE (5) to activate the team member's camera.

Table 2. Sample of the model: statechart behavior of the team leader (top), and LSC specification describes Flood task(bottom)

Samples of the Model	Description
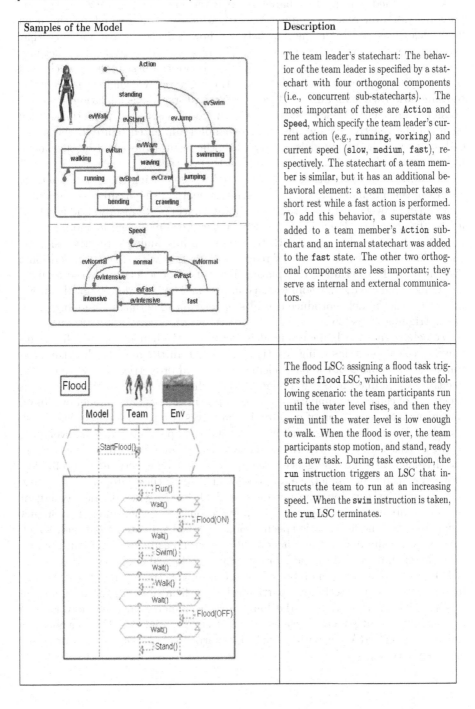	The team leader's statechart: The behavior of the team leader is specified by a statechart with four orthogonal components (i.e., concurrent sub-statecharts). The most important of these are Action and Speed, which specify the team leader's current action (e.g., running, working) and current speed (slow, medium, fast), respectively. The statechart of a team member is similar, but it has an additional behavioral element: a team member takes a short rest while a fast action is performed. To add this behavior, a superstate was added to a team member's Action sub-chart and an internal statechart was added to the fast state. The other two orthogonal components are less important; they serve as internal and external communicators.
	The flood LSC: assigning a flood task triggers the flood LSC, which initiates the following scenario: the team participants run until the water level rises, and then they swim until the water level is low enough to walk. When the flood is over, the team participants stop motion, and stand, ready for a new task. During task execution, the run instruction triggers an LSC that instructs the team to run at an increasing speed. When the swim instruction is taken, the run LSC terminates.

using the ROOM formalism. A similar approach was employed in [37] to model the Ethylene-Pathway in Arabidopsis thaliana using statecharts and LSCs.

As mentioned, the present paper is an extension and generalization of our previous work on reactive animation [10], which was motivated by the need for a clean way to bridge the gap between how objects behave and how that behavior should show up on the screen. The idea was to separate the reactivity from the visualization, making it possible to choose the best tools for each, and thus enjoying the benefits of both worlds. Having a reactive engine that controls the simulation while an animated front-end monitors the visualization, makes it possible to model large-scale systems with many objects and interactions. Each agent in the reactive engine has a corresponding animated figure in the front-end, and since the two models are separate, they can be designed by specialists in their particular fields using any state-of-the-art tools.

In [10], reactive animation was illustrated by simulating the development of T-cells in the Thymus gland, and it was implemented using a direct communication socket between a state-based model (statecharts in Rhapsody) and a 2D animated front-end (in Flash). The present paper provides a generic, modular and fully distributed architecture, with the ability to link multiple reactive engines, and it illustrates the feasibility of reactive animation with a three-dimensional visualization. Furthermore, this platform was used beneficially for the realistic modeling of pancreatic development, a complex and large-scale biological system. Moreover, the model reproduced results of relevant experiments and suggested new intriguing ideas [36].

We believe that our biological models emphasize the benefit for modeling complex large-scale systems using reactive animation. In the models, the visualized concurrent execution of the basic elements revealed properties that were not explicitly programmed into the model. Rather, they emerge from the concurrent execution of cells as a population. For example, in the pancreas model, we found that concurrent execution of pancreatic cells gives rise to a property that corresponds well with first transition clusters found to appear early in the developing organ *in vivo* [36]. Similarly, the concurrent execution of T-cell development in the thymus led to the emergence of competitive behavior between the cells [11]. Moreover, using the model, we analyzed and studied these properties and suggested some insights into the phenomena [8,11,36]. In general, since emergent properties are dynamic properties of a population, it is rather difficult to predict them from the model's static specifications. At the animated front-end, which visualizes the simulation, the phenomenon is often easily seen and can then be carefully examined against the literature for a biological explanation.

While we have used a number of examples to illustrate generic reactive animation, we find the technique particularly beneficial for large-scale biological systems. We envision that in the long run, the pancreas project may lead to new insights about pancreas-related diseases, such as diabetes. Furthermore, we feel that generic RA my perhaps help in efforts to build an *in-silico* organ or organism (see, e.g., [19]).

Acknowledgments. We would like to thank Shahar Maoz and Gera Weiss for their help and support in setting the architecture and designing the examples.

References

1. 3D Game Studio, http://www.3dgamestudio.com
2. Amir-Kroll, H., Sadot, A., Cohen, I.R., Harel, D.: GemCell: A Generic Platform for Modeling Multi-Cellular Biological Systems. T. Comp. Sys. Biology (to appear, 2007)
3. Axelrod, J.D.: Cell Shape in Proliferating Epithelia: A Multifaceted Problem. Cell 126, 643–645 (2006)
4. Barak, D., Harel, D., Marelly, R.: InterPlay: Horizontal Scale-Up and Transition to Design in Scenario-Based Programming. In: Desel, J., Reisig, W., Rozenberg, G. (eds.) ACPN 2003. LNCS, vol. 3098, pp. 66–86. Springer, Heidelberg (2004)
5. Brooks, R.A.: Elephants Don't Play Chess. Robotics and Autonomous Systems 6, 3–15 (1990)
6. Cardelli, L.: Abstract Machines of Systems Biology. T. Comp. Sys. Biology 3, 145–168 (2005)
7. Ciliberto, A., Novak, B., Tyson, J.J.: Mathematical Model of the Morphogenesis Checkpoint in Budding Yeast. J. Cell. Biol. 163, 1243–1254 (2003)
8. Cohen, I.R., Harel, D.: Explaining a Complex Living System: Dynamics, Multiscaling and Emergence. J. R. Soc. Interface 4, 175–182 (2007)
9. Damm, W., Harel, D.: LSCs: Breathing Life into Message Sequence Charts. Formal Methods in System Design 19, 45–80 (2001)
10. Efroni, S., Harel, D., Cohen, I.R.: Reactive Animation: Realistic Modeling of Complex Dynamic Systems. IEEE Computer 38, 38–47 (2005)
11. Efroni, S., Harel, D., Cohen, I.R.: Emergent Dynamics of Thymocyte Development and Lineage Determination. PLoS Comput. Biol. 3, 13 (2007)
12. Finkelstein, A., Hetherington, J., Li, L., Margoninski, O., Saffrey, P., Seymour, R., Warner, A.: Computational Challenges of Systems Biology. IEEE Computer 37, 26–33 (2004)
13. Fisher, J., Henzinger, T.A.: Executable cell biology. Nat. Biotechnol. 25, 1239–1249 (2007)
14. Ghosh, R., Tomlin, C.: Symbolic Reachable Set Computation of Piecewise Affine Hybrid Automata and its Application to Biological Modelling: Delta-Notch Protein Signalling. Syst. Biol (Stevenage) 1, 170–183 (2004)
15. Gibson, M.C., Patel, A.B., Nagpal, R., Perrimon, N.: The emergence of geometric order in proliferating metazoan epithelia. Nature 442, 1038–1041 (2006)
16. Harel, D.: Dynamic Logic. In: Gabbay, D., Guenthner, F. (eds.) Handbook of Philosophical Logic, vol. 2, pp. 497–604. Reidel, Dordrecht (1984)
17. Harel, D.: Statecharts: A Visual Formalism for Complex Systems. Sci. Comput. Programming 8, 231–274 (1987)
18. Harel, D.: A Grand Challenge for Computing: Full Reactive Modeling of a Multi-Cellular Animal. Bulletin of the EATCS 81, 226–235 (2003)
19. Harel, D.: A Turing-like test for biological modeling. Nat. Biotechnol. 23, 495–496 (2005)
20. Harel, D., Gery, E.: Executable Object Modeling with Statecharts. IEEE Computer 30, 31–42 (1997)

21. Harel, D., Kleinbort, A., Maoz, S.: S2A: A Compiler for Multi-modal UML Sequence Diagrams. In: Dwyer, M.B., Lopes, A. (eds.) FASE 2007. LNCS, vol. 4422, pp. 121–124. Springer, Heidelberg (2007)
22. Harel, D., Marelly, R.: Come, Let's Play: Scenario-Based Programming Using LSCs and the Play-Engine. Springer, Heidelberg (2003)
23. Harel, D., Pnueli, A.: On the Development of Reactive Systems. In: Apt, K.R. (ed.) Logics and Models of Concurrent Systems. NATO ASI Series, vol. F-13, pp. 477–498 (1985)
24. Heath, J., Kwiatkowska, M., Norman, G., Parker, D., Tymchyshyn, O.: Probabilistic Model Checking of Complex Biological Pathways. In: Priami, C. (ed.) CMSB 2006. LNCS (LNBI), vol. 4210, pp. 32–47. Springer, Heidelberg (2006)
25. Kam, N., Harel, D., Kugler, H., Marelly, R., Pnueli, A., Hubbard, E.J.A., Stern, M.: Formal Modeling of C. elegans Development: A Scenario-Based Approach. In: Priami, C. (ed.) CMSB 2003. LNCS, vol. 2602, pp. 4–20. Springer, Heidelberg (2003)
26. Kam, N., Kugler, H., Appleby, L., Pnueli, A., Harel, D., Stern, M.J., Hubbard, E.J.A.: A scenario-based approach to modeling development (I): Rationale, hypothesis testing, simulations and experimental follow-up (submitted, 2007)
27. Kugler, H., Kam, N., Marelly, R., Appleby, L., Pnueli, A., Harel, D., Stern, M.J., Hubbard, E.J.A.: A scenario-based approach to modeling development (II): A prototype model of C. elegans vulval cell fate specification (submitted, 2007)
28. The MathWorks, http://www.mathworks.com
29. Nelson, C.M., Vanduijn, M.M., Inman, J.L., Fletcher, D.A., Bissell, M.J.: Tissue geometry determines sites of mammary branching morphogenesis in organotypic cultures. Science 314, 298–300 (2006)
30. Noble, D.: The heart is already working. Biochem. Soc. Trans. 33, 539–542 (2005)
31. Priami, C., Quaglia, P.: Modelling the dynamics of biosystems. Briefings in Bioinformatics 5, 259–269 (2004)
32. Regev, A., Shapiro, E.: Cellular abstractions: Cells as computation. Nature 419, 343 (2002)
33. Regev, A., Silverman, W., Shapiro, E.: Representation and simulation of biochemical processes using the pi-calculus process algebra. In: Pacific Symposium on Biocomputing, pp. 459–470 (2001)
34. Roux-Rouquié, M., da Rosa, D.S.: Ten Top Reasons for Systems Biology to Get into Model-Driven Engineering. In: GaMMa 2006: Proc. of the 2006 international workshop on Global integrated model management, pp. 55–58 (2006)
35. Sadot, A., Fisher, J., Barak, D., Admanit, Y., Stern, M.J., Hubbard, E.J.A., Harel, D.: Towards Verified Biological Models. IEEE/ACM Trans. Comput. Biology and Bioinformatics (to appear, 2007)
36. Setty, Y., Cohen, I.R., Dor, Y., Harel, D.: Four-Dimensional Realistic Modeling of Pancreatic Organogenesis (submitted, 2007)
37. Taubner, C., Merker, T.: Discrete Modelling of the Ethylene-Pathway. In: ICDEW 2005: Proceedings of the 21st International Conference on Data Engineering Workshops, 1152 (2005)
38. Telelogic, http://www.telelogic.com
39. Webb, K., White, T.: Cell Modeling with Reusable Agent-based Formalisms. Applied Intelligence 24, 169–181 (2006)

Bounded Asynchrony: Concurrency for Modeling Cell-Cell Interactions [*]

Jasmin Fisher[1], Thomas A. Henzinger[2], Maria Mateescu[2], and Nir Piterman[3]

[1] Microsoft Research, Cambridge UK
[2] EPFL, Switzerland
[3] Imperial College London, UK

Abstract. We introduce *bounded asynchrony*, a notion of concurrency tailored to the modeling of biological cell-cell interactions. Bounded asynchrony is the result of a scheduler that bounds the number of steps that one process gets ahead of other processes; this allows the components of a system to move independently while keeping them coupled. Bounded asynchrony accurately reproduces the experimental observations made about certain cell-cell interactions: its constrained nondeterminism captures the variability observed in cells that, although equally potent, assume distinct fates. Real-life cells are not "scheduled", but we show that distributed real-time behavior can lead to component interactions that are observationally equivalent to bounded asynchrony; this provides a possible mechanistic explanation for the phenomena observed during cell fate specification.

We use model checking to determine cell fates. The nondeterminism of bounded asynchrony causes state explosion during model checking, but partial-order methods are not directly applicable. We present a new algorithm that reduces the number of states that need to be explored: our optimization takes advantage of the bounded-asynchronous progress and the spatially local interactions of components that model cells. We compare our own communication-based reduction with partial-order reduction (on a restricted form of bounded asynchrony) and experiments illustrate that our algorithm leads to significant savings.

1 Introduction

Computational modeling of biological systems is becoming increasingly important in efforts to better understand complex biological behaviors. In recent years, formal methods have been used to construct and analyze such biological models. The approach, dubbed "executable biology" [10], is becoming increasingly popular. Various formalisms are putting the executable biology framework into practice. For example, Petri-nets [3,7], process calculi [22,15], interacting state-machines [9,11], and hybrid automata [13,2]. In many cases, the analysis of these models includes reachability analysis and model checking in addition to traditional simulations.

This paper focuses on interacting state-machines as a tool for biological modeling [18,8,19,12,23,9,11]. This approach has recently led to various biological discoveries, and modeling works that were done using this approach have appeared in high impact biological journals [12,9,11]. These are discrete, state-based models that are used as high-level abstractions of biological systems' behavior.

[*] Supported in part by the Swiss National Science Foundation (grant 205321-111840).

J. Fisher (Ed.): FMSB 2008, LNBI 5054, pp. 17–32, 2008.
© Springer-Verlag Berlin Heidelberg 2008

```
var pathway,signal:{0..4};

pathway_atom
init
[] true -> path := 1;
update
[] (0<path<4) & no_input & next(signal)<4 -> path := path+1;
[] (0<path<4) & input & next(signal)<4 -> path := 4;
[] (0<path<4) & next(signal)=4 -> path := 0;

signal_atom
init
[] true -> signal := 3;
update
[] neighborpath=4 & signal>0 -> signal := 4;
[] neighborpath<4 & path=4 -> signal := 0;
[] neighborpath<4 & path<4 & 0<signal<4 -> signal := signal-1;
```

Fig. 1. Program for abstract model

When using interacting state-machine models to describe a biological behavior, we are facing the question of how to compose its components. We find that the two standard notions of concurrency (in this context), synchrony and asynchrony, are either too constrained or too loose when modeling certain biological behaviors such as cell-cell interactions. When we try to model cell-cell interactions, we find that synchronous composition is too rigid, making it impossible to break the symmetry between processes without the introduction of additional artificial mechanisms. On the other hand, asynchronous composition introduces a difficulty in deciding when to stop waiting for a signal that may never arrive, again requiring artificial mechanisms.[1]

Biological motivation. We further explain why the standard notions of concurrency may be inappropriate for modeling certain biological processes. We give a model representing very abstractly a race between two processes in adjacent cells that assume two different cell fates. The fate a cell chooses depends on two proteins, denoted *pathway* and *signal*, below. The pathway encourages the cell to adopt fate1 while the signal encourages the cell to adopt fate2. In the process we are interested in, pathway starts increasing slowly. When pathway reaches a certain level, it forces the cell to adopt fate1. At the same time, pathway encourages the signal in neighbor cells to increase and inhibits the pathway in the neighbor cell. The signal starts in some low level and if not encouraged goes down and vanishes. If, however, it is encouraged, it goes up, inhibiting the pathway in the same cell, and causing the cell to adopt fate2. A simple model reproducing this behavior is given in Fig. 1.

---

[1] We treat here biological processes as computer processes. For example, when we say 'waiting', 'message', or 'decide' we relate to biological processes that take time to complete, and if allowed to continue undisturbed may lead to irreversible consequences. Thus, as long as the process is going on the system 'waits', and if the process is not disturbed ('does not receive a message'), it 'decides'.

We are interested in three behaviors. First, when a cell is run in isolation, the pathway should prevail and the cell should assume fate1. Second, when two cells run in parallel either of them can get fate1 and the other fate2. There are also rare cases where both cells assume fate1. Third, when one of the cells gets an external boost to the pathway it is always the case that this cell adopts fate1 and the other fate2.

Already this simplified model explains the problems with the normal notions of concurrency. In order to allow for the second behavior we have to break the symmetry between the cells. This suggests that some form of asynchrony is appropriate. The combination of the first and third behaviors shows that the asynchronicity has to be bounded. Indeed, in an asynchronous setting a process cannot distinguish between the case that it is alone and the case that the scheduler chooses it over other processes for a long time.

Although very simple, this model is akin to many biological processes in different species. For example, a similar process occurs during the formation of the wing of the Drosophila fruit fly [13]. Ghosh and Tomlin's work provides a detailed model (using hybrid automata) of this process. The formation of the C. elegans vulva also includes a similar process [11]. Our model of C. elegans vulval development uses the notion of bounded asynchrony. Using bounded asynchrony we separate the modeling environment from the model itself and suggest biological insights that were validated experimentally [11].

Formal modeling: bounded asynchrony. For this reason, we introduce a notion of *bounded asynchrony* into our biological models, which allows components of a biological system to proceed approximately along the same time-line. In order to implement bounded asynchrony, we associate a rate with every process. The rate determines the time t that the process takes to complete an action. A process that works according to rate t performs, in the long run, one action every tth round. This way, processes that work according to the same rate work more or less concurrently, and are always at the same stage of computation, however, the action itself can be taken first by either one of the processes or concurrently, and the order may change from round to round[2]. Other notions of bounded asynchrony either permit processes to 'drift apart', allowing one process to take arbitrarily more actions than another process, or do not generalize naturally to processes working according to different rates.

Having the above mentioned example in mind, we define the notion of bounded asynchrony by introducing an explicit scheduler that instructs each of the cells when it is allowed to move. Thus, our system is in fact a synchronous system with a non-deterministic scheduler instructing which processes to move when. We find this notion of bounded asynchrony consistent with the observations made in cell-cell interactions. As explained, asynchrony is essential in order to break the symmetry between cells (processes). It is important to separate the biological mechanism from the synchronization mechanism, otherwise the model seems removed from the biology. On the other hand, much like in distributed protocols, a process has to know when to give up on waiting for messages that do not arrive. With classical asynchrony this is impossible and we are forced to add some synchronizing mechanism. Again, in the context of biology, such a mechanism should be presented in terms of the modeling environment. When introducing bounded asynchrony both problems are solved. The asynchrony breaks the

---

[2] We note that this process is not memoryless, making continuous time Markov chains inappropriate. This issue is discussed further below.

symmetry and the bound allows processes to decide when to stop waiting. In addition, the asynchrony introduces limited nondeterminism that captures the diversity of results often observed in biology.

Possible mechanistic explanation: real time. In some cases, biological systems allow central synchronization. For example, during animal development, it may happen that several cells are arrested in some state until some external signal tells all of them to advance. However, these synchronization mechanisms operate on a larger scale and over time periods that are much longer than the events described by our model. Thus, we do not believe that there is a centralized scheduler that instructs the processes when to move. The behaviors we describe are observed in practice, suggesting that there is some mechanism that actually makes the system work this way. This mechanism has to be distributed between the cells. We show that bounded asynchrony arises as a natural abstraction of a specific type of clocked transition systems, where each component has an internal clock. This suggests that similar ideas may be used for the abstraction of certain types of real time systems. Of less importance here, it also may be related to the actual mechanism that creates the emergent property of bounded asynchrony.

Model checking: scheduler optimization. The scheduler we introduce to define bounded asynchrony consists of adding variables that memorize which of the processes has already performed an action in the current round. When we come to analyze such a system we find that, much like in asynchronous systems, many different choices of the scheduler lead to the same states. Motivated by partial-order reduction [6], we show that in some cases only part of the interleavings need to be explored. Specifically, our method applies in configurations of the system where communication is locally restricted. In such cases, we can suggest alternative schedulers that explore only a fraction of the possible interleavings, however, explore all possible computations of the system. We also compare our techniques with partial-order reduction in a restricted setting with no concurrent moves. Experimental evaluation shows that our techniques lead to significant improvement. We are not familiar with works that analyze the structure of communication in a specific concurrent system and use this structure to improve model checking.

Related (and unrelated) models. The comparison of such abstract models with the more detailed differential equations or stochastic process calculi models is a fascinating subject, however, this is not the focus of this paper. Here, we assume that both approaches can suggest helpful insights to biology. We are also not interested in a particular biological model but rather in advancing the computer science theory supporting the construction of abstract biological models.

There are mainly two approaches to handle concurrency in abstract biological models. One prevalent approach is to create a continuous time Markov chain (CTMC). This approach is usually used with models that aim to capture molecular interactions [14,22]. Then, the set of enabled reactions compete according to a continuous probability rate (usually, the χ-distribution). Once one reaction has occurred, a new set of enabled reactions is computed, and the process repeats. This kind of model requires exact quantitative data regarding number of molecules and reaction rates. Such accurate data is sometimes hard to obtain; indeed, even the data as to exactly which molecules are involved in the process may be missing (as is the case in the *C. elegans* model). Our models are very far from the molecular level, they are very abstract, and scheduler choices are made on the cellular level. When considering processes abstractly the scheduling is no longer memoryless, making CTMCs inappropriate. For example, consider a CTMC obtained from our model in Fig. 1 by setting two cells in motion according to the same rate. Consider the experiment where one of the cells is getting a boost to its pathway. The probability of the other cell performing 4 consecutive actions (which would lead to it getting fate1) is $\frac{1}{16}$, while this cannot occur in the real system. In addition, the probability of both cells assuming fate1 is 0, as the cells cannot move simultaneously.

A different approach, common in *Boolean networks* [20,4,5], is to use asynchrony between the substances. Again, this approach is usually applied to models that aim to capture molecular interactions, however, in an abstract way. Asynchronous updates of the different components is used as an over-approximation of the actual updates. If the system satisfies its requirements under asynchronous composition, it clearly satisfies them under more restricted compositions. We note, however, that these models are used primarily to analyze the steady-state behavior of models (i.e., loops that have no outgoing edges). As asynchrony over-approximates the required composition, such steady-state attractors are attractors also in more restricted compositions, justifying this kind of analysis. For our needs, we find unbounded asynchrony inappropriate.

Bounded asynchrony is in a sense the dual of GALS (globally-asynchronous-locally-synchronous): it represents systems that look globally, viewed at a coarse time granularity, essentially synchronous, while they behave locally asynchronous, at a finer time granularity. Efficient implementations of synchronous embedded architectures also fall into this category. For example, time-triggered languages such as Giotto [16] have a synchronous semantics, yet may be implemented using a variety of different scheduling and communication protocols.

2 Modeling Cell-Cell Interactions

Here we describe in brief the vulval development of the earthworm *C. elegans*, a process that is similar to the process described in Section 1. In a separate paper, intended for a biological audience, we describe a computational model of this process [11]. Our model is of abstraction level similar to the example in Section 1. The paper mainly covers the biological insights the model suggests and their biological validation, bounded asynchrony and its algorithmic aspects are not covered. Readers interested in the model itself are referred to [1]

The *C. elegans* vulva (the egg-laying system) normally derives from three vulval precursor cells (VPCs) that are members of a larger set of six VPCs, named P3.p – P8.p (see Fig. 2). Each of the six VPCs is multipotent, capable of adopting one of three cell fates (1°, 2°, or 3°). The actual fate each cell adopts depends upon three intercellular signals: the epidermal growth factor receptor (EGFR) inductive signal emanating from the gonadal anchor cell (AC), the LIN-12/Notch lateral signal operating between VPCs, and the inhibitory signal coming from the surrounding hypodermal syncytium (Fig. 2). VPC fates in wild-type animals are influenced by their distance from the AC: the cell closest to the AC (P6.p) becomes 1°, the next closest (P5.p and P7.p) become 2°, and the most distant cells (P3.p, P4.p, and P8.p) become 3°.

The postulated mechanism that drives VPC fate specification is as follows. The EGFR activates an internal cascade that consists of a few proteins. An activation of

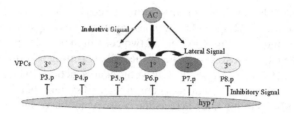

Fig. 2. The intercellular signaling specifying three cell fates during *C. elegans* vulval development

this cascade starts by an increase in the level of the first protein. An increase in the level of a protein causes the next in line to increase as well until full activation of the cascade. There is a signal emanating from the dermis of the worm inhibiting this cascade (i.e., keeping it inactive). In the absence of this signal the cascade is initiated without an external stimulus. The fate specification process starts when the anchor cell (AC) sends an inductive signal (IS) to the VPCs. In a VPC receiving a low level of IS, the cascade is kept inactive and the VPC adopts a 3° fate. A high IS causes the cascade to initiate. The cascade also causes a lateral signal (LS) to be sent to the immediate surrounding of the cell only after disabling the receptors to this signal in the same cell. Finally, the full activation of the cascade causes the cell to adopt a 1° fate. A medium level of IS also causes the cascade to initiate, however slower. If the cascade is fully activated, the cell assumes a 1° fate. The perception of the lateral signal sent by a neighbor cell turns off the cascade and causes the cell to assume a 2° fate. By experimenting with the system (creating mutations and perturbing the system in various ways) certain signals / receptors can be shut-down or encouraged, and the behavior of the system is observed in these cases. A model should be able to reproduce the behaviors observed in actual experiments.

The examples that show the importance of bounded asynchrony in this context are very similar to the example in Section 1. Consider the following three experiments. First, in the case that the inhibitory signal is shut-down, all cells initiate the protein cascade, then some cells manage to stop the cascade in their neighbors. These cells adopt a 1° fate and their neighbors a 2° fate. In addition, in repeating this experiment, different cells assume a 1° fate. This suggests that we have to introduce asynchrony in order to break the symmetry between the cells. In the case that the receptor of the lateral signal (LS) is shut-down, cells that initiate the cascade do not perceive the LS and assume a 1° fate. This suggests that in the case that the cascade initiates, then at some point, it should decide that the LS does not arrive. Finally, in a normal (unperturbed) system it is always the case that if the cascade initiates slowly then it is counteracted by the LS. However, if we use asynchrony in the unperturbed system the scheduler may decide to delay the LS arbitrarily long; and the cells in which the cascade is initiated slowly would assume a 1° fate.

It is relatively simple to see that when using bounded asynchrony, a cell that awaits an LS knows when to stop waiting. It can decide that the LS is not going to arrive at all and safely continue with its computation. Bounded asynchrony introduces small races between processes. In our case, cells proceed towards fate acquisition, and a cell that adopts a 1° fate inhibits its neighbors from assuming the same fate. If a cell moves slightly faster than its neighbors, it gets to assume a $1°$ fate and inhibits its neighbors that assume a $2°$ fate. Dually, if a cell moves slightly slower than its neighbors, the neighbors get to assume the $1°$ fate and inhibit the cell that assumes a $2°$ fate. This reproduces very nicely the unstable fate patterns (the same cell assumes different fates in repetitions of the same experiment) observed in experiments involving the inhibitory signal shut-down.

It is an interesting choice whether to allow processes to move concurrently or not. Intuitively, we view concurrent moves of processes as the mirror of the rare event where the difference between biological processes is so small that it is ignored. We allow processes to move concurrently, creating situations in which cells proceed synchronously performing the same sequence of actions (in particular assuming the same fate). In some mutations leading to multiple possible outcomes, we know that some of the outcomes are rarely observed. When we disallow concurrent moves of processes, these rare observations disappear from the model, matching our intuition of bounded asynchrony.

3 Bounded Asynchrony

In this section we define the notion of bounded asynchrony. We first define transition systems and then proceed to the definition of bounded asynchrony.

3.1 Transition Systems

A *transition system* (TS) $\mathcal{D} = \langle V, W, \Theta, \rho \rangle$ consists of the following components.

- $V = \{u_1, \dots, u_n\}$: A finite set of typed *state variables* over finite domains. We define a *state* s to be a type-consistent interpretation of V, assigning to each variable $u \in V$ a value $s[u]$ in its domain. We denote by Σ the set of all states. For an assertion φ, we say that s is a φ-state if $s \models \varphi$.
- $W \subseteq V$: A set of *owned* variables. These are the variables that only $\mathcal{D}$ may change. The set W includes the Boolean *scheduling variable* a.
- Θ : The *initial condition*. This is an assertion characterizing all the initial states of the TS. A state is called *initial* if it satisfies Θ.
- ρ: A *transition relation*. This is an assertion $\rho(V, V')$, relating a state $s \in \Sigma$ to its $\mathcal{D}$-successor $s' \in \Sigma$ by referring to both unprimed and primed versions of the state variables. The transition relation $\rho(V, V')$ identifies state s' as a $\mathcal{D}$-*successor* of state s if $(s, s') \models \rho(V, V')$. The transition relation ρ has the form $(a \neq a' \land \rho') \lor (W = W')$, where a is the scheduling variable. In what follows we restrict our attention to systems that use a scheduling variable.

A *run* of $\mathcal{D}$ is a sequence of states $\sigma : s_0, s_1, \dots$, satisfying the requirements of (a) *Initiality:* s_0 is initial, i.e., $s_0 \models \Theta$; (b) *Consecution:* for every $j \geq 0$, the state s_{j+1} is a $\mathcal{D}$-successor of the state s_j. We denote by $runs(\mathcal{D})$ the set of runs of $\mathcal{D}$. We can divide the run to transitions where $\mathcal{D}$ stutters (i.e., a and all variables in W do not change) and where $\mathcal{D}$ moves (i.e., a flips its value and variables in W may change).

Given systems $\mathcal{D}_1 : \langle V_1, W_1, \Theta_1, \rho_1 \rangle$ and $\mathcal{D}_2 : \langle V_2, W_2, \Theta_2, \rho_2 \rangle$ such that $W_1 \cap W_2 = \emptyset$, the *parallel composition*, denoted by $\mathcal{D}_1 \| \mathcal{D}_2$, is the TS $\langle V, W, \Theta, \rho \rangle$ where $V = V_1 \cup V_2$, $W = W_1 \cup W_2 \cup \{a\}$, $\Theta = \Theta_1 \land \Theta_2$, and $\rho = \rho_1 \land \rho_2 \land \rho'$, the variable

a is the scheduling variable of $\mathcal{D}_1 \parallel \mathcal{D}_2$ and ρ' is as follows[3]

$$\rho' = (a \neq a') \iff [(a_1 \neq a_1') \vee (a_2 \neq a_2')]$$

For more details, we refer the reader to [21].

The *projection* of a state s on a set $V' \subseteq V$, denoted $s\Downarrow_{V'}$, is the interpretation of the variables in V' according to their values in s. Projection is generalized to sequences of states and to sets of sequences of states in the natural way.

3.2   Explicit Scheduler

We define bounded asynchrony by supplying an explicit scheduler that lets all processes proceed asynchronously, however, does not permit any process to proceed faster than other processes. Intuitively, the system has one macro-step in which each of the processes performs one micro-step (or sometimes none), keeping all processes together (regarding the number of actions). The order of actions between the subprocesses is completely non-deterministic. Thus, some of the processes may move together and some one after the other. We start with a scheduler that allows all processes to proceed according to the same rate. We then explain how to generalize to a scheduler that implements bounded asynchrony between processes with different rates.

We start by considering a set of processes all working according to the same rate (without loss of generality the rate is 1). In this case, the resulting behavior is that every process does one micro-step in every macro-step of the system. Namely, we can choose a subset of the processes, let them take a move, then continue with the remaining processes until completing one macro-step. We create a TS that schedules actions accordingly. The scheduler has a Boolean variable b_i associated with every process P_i. A move of P_i is forced when b_i changes from false to true. Once all b_is are set to true, they are all set concurrently to false (and no process moves).

More formally, consider n TSs $P_1, \ldots, P_n$. For $1 \leq i \leq n$, let a_i be the scheduling variable of P_i and let $(\rho_i \wedge a_i \neq a_i') \vee (W_i = W_i')$ be the transition relation of P_i. We define a scheduler $S = \langle V, W, \Theta, \rho \rangle$, where $V = W = \{b_1, \ldots, b_n\}$ and b_i is Boolean for all $1 \leq i \leq n$, $\Theta = \bigwedge_{i=1}^{n} \overline{b_i}$, and ρ is defined as follows:

$$\rho = \left(\Omega \rightarrow \bigwedge_{i=1}^{n} (b_i \rightarrow b_i') \quad \wedge \quad \overline{\Omega} \rightarrow \bigwedge_{i=1}^{n} \overline{b_i'} \right) \tag{1}$$

Where $\Omega = \bigvee_{i=1}^{n} \overline{b_i}$ denotes the assertion that at least one variable b_i is still false.

The *bounded asynchronous parallel composition* of $P_1, \ldots, P_n$ according to the rate 1, denoted $P_1^1 \parallel_{ba} \cdots \parallel_{ba} P_n^1$, is $S \parallel P_1 \parallel \cdots \parallel P_n$ with the following additional conjunct added to the transition:

$$\bigwedge_{i=1}^{n} (a_i \neq a_i' \iff (\overline{b_i} \wedge b_i')) \tag{2}$$

Thus, the scheduling variable of P_i is forced to change when b_i is set to true.

---

[3] Notice that, in the case that $\mathcal{D}_1$ and $\mathcal{D}_2$ have stutter transitions, this composition is neither synchronous nor asynchronous in the classical sense.

We consider now the more general case of processes working with general rates. In this case, we use the same system of Boolean variables but in addition have a counter that counts the number of steps. A process is allowed to make a move only when its rate divides the value of the counter. More formally, let the rates of $P_1, \ldots, P_n$ be $t_1, \ldots, t_n$. For $1 \leq i \leq n$, let a_i be the scheduling variable of P_i and let $(\rho_i \wedge a_i \neq a_i') \vee (W_i = W_i')$ be the transition relation of P_i. We define a scheduler $S = \langle V, W, \Theta, \rho \rangle$ with the following components:

- $V = W = \{b_1, \ldots, b_n, c\}$. Forall i we have b_i is Boolean, and c ranges over $\{1, \ldots, lcm(t_1, \ldots, t_n)\}$, where lcm is the least common multiplier.
- $\Theta = (c=1) \wedge \bigwedge_{i=1}^{n} \overline{b_i}$.
- Let $\Omega = \bigvee_{i=1}^{n} (\overline{b_i} \wedge (c \bmod t_i = 0))$ denote the assertion that at least one variable b_i for which the rate t_i divides the counter is still false.

$$\rho = (\Omega \to \bigwedge_{i=1}^{n}(b_i \to b_i') \wedge (c=c')) \wedge \left(\overline{\Omega} \to \bigwedge_{i=1}^{n}(\overline{b_i'} \wedge (c'=c \oplus 1)) \right) \wedge \\ \left(\bigwedge_{i=1}^{n}((\overline{b_i} \wedge (c \bmod t_i \neq 0)) \to \overline{b_i'}) \right) \tag{3}$$

The *bounded asynchronous parallel composition* of $P_1, \ldots, P_n$ according to rates $t_1,$ $\ldots, t_n$, denoted $P_1^{t_1} \|_{ba} \cdots \|_{ba} P_n^{t_n}$, is $S \| P_1 \| \cdots \| P_n$ with the the the conjunct in Equation (2) added to the transition.

We note that there are many possible ways to implement this restriction of the possible interleavings between processes. Essentially, they all boil down to counting the number of moves made by each process and allowing / disallowing processes to move according to the values of counters.

4   Model Checking

Partial Order Reduction (POR) [6] is a technique that takes advantage of the fact that in asynchronous systems many interleavings lead to the same results. It does this by not exploring some redundant interleavings, more accurately, by shrinking the set of successors of a state while preserving system behavior. Existing algorithms are designed for (unbounded) asynchronous systems and do not directly adapt to our kind of models (see below). Although, at the moment, we are unable to suggest POR techniques for bounded asynchrony, we propose an algorithm that exploits the restricted communication encountered in systems that model cell-cell interaction, we refer to our algorithm as *communication based reduction*, or CBR for short. Like POR, our algorithm searches only some of the possible interleavings. For every interleaving, our algorithm explores an interleaving that visits the same states on a macro-step level. We reduce the reachable region of the scheduler from exponential size to polynomial size in the number of processes, and thus we have a direct and important impact on enumerative model checking. Our approach is applicable to all linear time properties whose validity is preserved by restricting attention to macro steps. Much like POR, the *next* operator cannot be handled. In particular, every property that relates to a single process (without next), and Boolean combinations of such properties, retain their validity.

4.1 Communication Based Reduction

The explicit scheduler S defined in Subsection 3.2 allows all possible interleavings of processes within a macro-step. We prove that we can construct a new scheduler that preserves system macro behavior (macro-step level behavior) but allows fewer interleavings. Let $\mathcal{P} = P_1^1 \parallel_{ba} \cdots \parallel_{ba} P_n^1$ be the bounded asynchronous composition of $P_1, P_2, \ldots, P_n$ according to rate 1 (see Section 3).[4]

We first formally define a *macro-step g* of $\mathcal{P}$ as a sequence of states $g : s = s_0, s_1, \ldots, s_m$ satisfying:

- g is a subsequence of a run,
- s_0 is initial with respect to the scheduler, i.e., $\overline{s_0[b_k]}$ holds for all $0 \leq k \leq n$,
- s_m is final with respect to the scheduler, i.e., $s_m[b_k]$ holds for all $0 \leq k \leq n$,
- s_m is the only final state in g.

A macro-step induces a total and a partial order over the processes of $\mathcal{P}$. The total order represents the order in which the processes move and we refer to it as the macro-step's interleaving. The partial order represents the order in which processes pass messages (via variables) and we refer to it as the macro-steps channel configuration.

Consider a macro-step $g : s = s_0, s_1, \ldots, s_m$ of $\mathcal{P}$. The *interleaving* of g, denoted $\mathcal{I}_g = (<_{Ig}, =_{Ig})$, is an order such that: $(P_k <_{Ig} P_l)$ if there exists s_i in g such that $s_i[b_k]\overline{s_i[b_l]}$ and $(P_k =_{Ig} P_l)$ if $(P_k \not<_{Ig} P_l) \wedge (P_l \not<_{Ig} P_k)$. That is, $(P_k <_{Ig} P_l)$ if P_k moves before P_l in the interleaving g.

We say that there is a communication channel c_{kl} connecting P_k and P_l if $V_k \cap V_l \neq \emptyset$. The *neighbor order* of g, denoted $(<_{Ng}, =_{Ng})$, is the partial order defined as the restriction of the interleaving of g to the neighboring processes. $P_k <_{Ng} P_l$ iff $P_k <_{Ig} P_l$ and there exists a channel c_{kl}. We define in a similar way $=_{Ng}$. The *channel configuration* of g, denoted $(<_{Cg}, =_{Cg})$, is the transitive closure of the neighbor order. That is, $P_k <_{Cg} P_l$ if a change in value of a variable of P_k in interleaving g can be sensed by P_l in the same interleaving.

Given a macro-step g, a channel c_{kl} may have one of three states: enabled from k to l, if $P_k <_{Cg} P_l$, enabled from l to k, if $P_l <_{Cg} P_k$, disabled, if $P_l =_{Cg} P_l$. Intuitively, a channel is enabled if it may propagate a value generated in the current macro-step.

Two interleavings are $\mathcal{P}$-*equivalent* if they induce the same channel configuration.

Within $\mathcal{P}$, we say that t is a *macro-successor* of s with respect to interleaving $\mathcal{I}$ if there exists a macro-step g with initial state s, interleaving $\mathcal{I}$ and final state t.

The following lemma establishes that two equivalent interleavings have the same set of macro-successors.

Lemma 1. *Consider two $\mathcal{P}$-equivalent interleavings $\mathcal{I}$ and $\mathcal{I}'$. If s' is a macro-successor of s with respect to $\mathcal{I}$, then s' is a macro-successor of s with respect to $\mathcal{I}'$.*

A scheduler that allows only one of two $\mathcal{P}$-equivalent interleavings preserves system macro-behavior. It follows that a scheduler that generates only one interleaving per channel configuration produces a correct macro-state behavior.

Here after we focus on the case of *line communication scheme* ($V_k \cap V_l = \emptyset$, for all $l \notin \{k - 1, k + 1\}$, $k \in (1..n)$). This is a common configuration in biological models

---

[4] Here, we only describe the case of processes running at equal rates. The same ideas can be easily extended to general rates.

where communication is very local. Extension to 2-dimensional configurations follows similar ideas.

Let c_k denote the channel $c_{k,k+1}$. In interleaving g, channel c_k is *enabled-right* if enabled from k to $k+1$, *enabled-left* if enabled from $k+1$ to k, and disabled as before.

Given a channel configuration we construct one interleaving that preserves it. Let $c_{r_0}, c_{r_1}, \ldots, c_{r_{mr}}$ be the right-enabled channels. Process P_{r_0} is oblivious to whatever happens in the same macro step in processes $P_{r_0+1}, \ldots, P_n$ because its communication with these processes happens through process P_{r_0+1} which moves after it. Thus, whatever actions are performed by processes $P_{r_0+1}, \ldots, P_n$ they do not affect the actions of processes $P_1, \ldots, P_{r_0}$. We may shuffle all the actions of processes $P_1, \ldots, P_{r_0}$ to the beginning of the interleaving preserving the right-enabled channel c_{r_0}. The new interleaving starts by handling all processes $P_1, \ldots, P_{r_0}$ from right to left. Let $c_{l_0}, c_{l_1}, \ldots, c_{l_{ml}}$ be the left-enabled channels in $1, \ldots, r_0$. Then, the order of moves is: first processes $P_{l_m+1}, \ldots, P_{r_0}$, then $P_{l_{m-1}+1}, \ldots, P_{l_m}$, and so on until $P_0, \ldots P_{l_1}$.

Next, using the same reasoning, we can handle the processes in the range $r_0 + 1, \ldots, r_1$ from right to left according to the left-enabled channels, and so on.

The CBR scheduler also uses the Boolean variables $b_1, \ldots, b_n$, however, the possible assignments are those where the processes can be partitioned to at most four maximal groups of consecutive processes that have either moved or not. More formally, we denote the value of $b_1, \ldots, b_n$ by a sequence of 0 and 1, then the configurations can be described by the following regular expressions: $0^+1^+0^+$, 1^+0^+, $1^+0^+1^+0^+$, and 1^+. With similar intuition configurations of the form 0^+, 0^+1^+, and $1^+0^+1^+$ are also reachable. For example, in a system with 6 processes the configurations 000111 and 110110 are reachable while the configuration 010101 is not. There are only $O(n^3)$ such reachable states, compared to 2^n reachable states in the original scheduler. Fig. 3(a) compares the number of states and transitions of the two schedulers (none, the scheduler described in Section 3 with no reduction vs. CBR, the scheduler described above) for different number of processes.

4.2 Experimental Evaluation

We compare experimentally the performance of the CBR scheduler with POR methods. We translate the model in Fig. 1 to Promela and use Spin [17] for a thorough analysis of the behavior of CBR.

We explain, intuitively, why POR is inappropriate for bounded asynchrony. We assume basic familiarity with POR. First, we find it very important that processes may move concurrently under bounded asynchrony. POR is developed for 'classical' asynchronous systems, thus, it does not allow for processes to move concurrently. Second, a macro-step in bounded asynchrony is a sequence of at most n local steps, and noticing that one interleaving is redundant may require exploration of more than 1 lookahead. Let us further explore this with an example. Suppose that we give up on concurrent moves and would like to use POR for reasoning about the same bounded asynchronous system. That is, processes in a line configuration where only neighbor processes may communicate. In the beginning of a macro-step, all processes are enabled. Communication between processes implies that we cannot find independent processes (such that the order of scheduling them does not matter), and we have to explore all possible n processes as the first process to move. With one process ahead of others, it is clear that

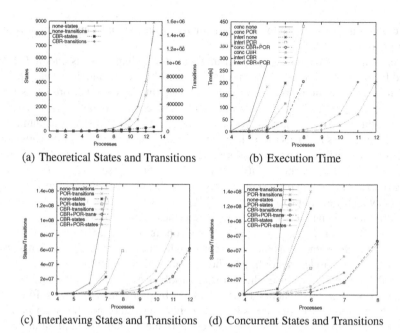

(a) Theoretical States and Transitions   (b) Execution Time

(c) Interleaving States and Transitions   (d) Concurrent States and Transitions

Fig. 3. Comparing theory and practice

the processes to the left of this process and to its right are no longer connected and the order between scheduling every process on the left and every process on the right can be exchanged. However, among the processes on one side, there is still dependency and the same selection by the scheduler has to be applied recursively. Overall, the number of possible interleavings to be checked is still exponential in n.[5] As exhibited by our experiments, POR does offer some reduction, however, this cannot be compared to the order of magnitude saving offered by using communication-based reduction.

We consider the bounded asynchronous composition of n cells in a line configuration. All processes start from the same state. If we disallow concurrent moves, we verify that there are no adjacent cells that assume fate1 (see Fig. 1). We add a mechanism that allows us to model concurrent moves using Spin's interleaving semantics. This mechanism consists of deciding to store the next values of variables in a local copy, allowing other processes to perform a computation according to the old values, and finally updating the new values. Obviously, this mechanism increases considerably the number of states in the system. For this case we verify that a cell assumes fate2 only if it has a neighbor that assumes fate1. We evaluate the CBR scheduler by considering the time for enumerative model checking and the number of states and transitions explored during model checking. We compare the behavior of the CBR scheduler with the basic scheduler described in Section 3 (simple scheduler) when POR is enabled and disabled. We perform two sets of experiments, both using Spin. The first set of experiments uses

---

[5] More accurately, the analysis is as follows. The number of interleavings of one process is $f(1) = 1$, the number of interleavings of zero processes is $f(0) = 0$. Generally, $f(n) = \Sigma_{i=1}^{n}(f(i-1) + f(n-i)) = 2f(n-1) + f(n-1) = 3f(n-1)$ and $f(n) = 2 \cdot 3^{n-2}$.

the normal interleaving semantics of Spin. In this case the size of the CBR scheduler is reduced from $O(n^3)$ to $O(n^2)$ states. This set of experiments includes running the simple scheduler without any reductions (none), the simple scheduler with POR (POR), the CBR scheduler (CBR), and the CBR scheduler with POR (CBR+POR). The second set of experiments includes a mechanism that makes Spin mimic the possibility of concurrent moves. We note that this additional mechanism increases the size of each process and that in order to communicate with the CBR scheduler each process has additional variables. Thus, the experiment is unfair with respect to the CBR scheduler. As before, this set of experiments includes running the simple scheduler (conc none), simple scheduler with POR (conc POR), CBR scheduler (conc CBR), and CBR scheduler with POR (conc CBR+POR). In all experiments, increasing the number of processes by one leads to memory overflow (10GB). For example, for the experiment with 9 processes, with the simple scheduler where POR is enabled, Spin requires more than 10GB of memory. Fig. 3(b) compares the model-checking time for the different experiments. Figures 3(c) and 3(d) compare the numbers of states and transitions explored in the first (interleaving semantics) and second (with mechanism mimicking concurrent moves) sets of experiments, respectively. For better scaling, the range of values covered by these figures does not include the number of transitions for the none-experiments in the cases of 7 and 5 processes, respectively. Notice that the size of the system itself increases exponentially with the number of processes. The experiments confirm that POR offers some improvement while the communication-based reduction affords a significant improvement when compared with the simple scheduler with POR.

The success of CBR in the context of bounded asynchrony suggests that it may be useful to analyze the communication structure in systems prior to model checking and to apply specific optimizations based on this analysis. Further research in this direction is out of the scope of this paper.

5   A Possible Mechanistic Explanation for Bounded Asynchrony

It is rather obvious that a scheduler such as the one we describe in Section 3 does not exist in real biological systems. While trying to describe biological behavior (of this type) in high-level requires us to use a notion like bounded asynchrony, it is not clear what is responsible for this kind of behavior in real systems. Obviously, no centralized control exists in this case, and there has to be some distributed mechanism that creates this kind of behavior. In this section we show that bounded asynchrony can be naturally used to abstract a special kind of distributed real-time mechanism. Thus, in some cases, similar scheduling mechanisms can be used to construct rough abstractions of real-time systems. From a biological point of view, it is an interesting challenge to design biological experiments that will confirm or falsify the hypothesis that internal clock-like mechanisms are responsible for the emerging behavior of bounded asynchrony.

We suggest clocked transition systems (CTS) as a possible distributed mechanism that produces bounded asynchrony. The systems we consider use a single clock, perform actions when this clock reaches a certain value, and reset the clock. We give a high-level description of the CTS we have in mind.

Consider the CTS Φ depicted in Fig. 4. The CTS has two Boolean variables s and a_p and one clock x_p. The values of s correspond to the two states in the figure. The CTS

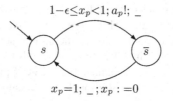

Fig. 4. CTS for one rate

is allowed to move from s to $\bar{s}$ when the clock x is in the range $[1 - \epsilon, 1)$, for some ϵ. When the CTS moves from $\bar{s}$ to s, it resets the clock back to 0. The variable a_p is the scheduling variable that this CTS sets; it changes when the system moves from s to $\bar{s}$, and does not change when the system moves from $\bar{s}$ to s. The possible computations of this system include the clock progressing until some point in $[1 - \epsilon, 1)$, then the system makes a transition from s to $\bar{s}$ while changing a_p, then the clock progresses until it is 1, and finally the system makes a transition from $\bar{s}$ to s. Then, the process repeats itself when the global time is $[2 - \epsilon, 2)$, $[3 - \epsilon, 3)$, and in general $[i - \epsilon, i)$ for every i.

Consider now the composition of Φ with a TS P that uses a_p as its scheduling variable. The composition of the two is a CTS in which moves of the TS P happen in the time range $[i-\epsilon, i)$ for every $i \in \mathbb{N}$. Suppose that we have two TS P and Q with scheduling variables a_p and a_q, respectively. We take the composition of two CTS as above using clocks x_p and x_q and the variables a_p and a_q. It follows that P and Q take approximately one time unit to make one move. However, the exact timing is not set. In a run of the system combined of the four CTSs the order of actions between P and Q is not determined. Every possible ordering of the actions is possible. In addition, the transitions that reset the clocks x_p and x_q ensure that the two TSs stay coupled. However long the execution, it cannot be the case that P takes significantly more actions than Q (in this case more than one). Under appropriate projection, the sequence of actions taken by the composition of the four systems, is equivalent to the sequence of actions taken by the bounded-asynchronous composition of P and Q with rate 1.

We now turn to consider the more general scheduler. Consider the CTSs in Fig. 5. They resemble the simple CTS presented above, however use the bounds of t_1 and t_2 time units, respectively. Denote the CTS using bound t_1 by Φ_1, and the CTS using bound t_2 by Φ_2. A computation of Φ_1 is a sequence of steps where time progresses until the range $[i \cdot t_1 - \epsilon, i \cdot t_1)$, then the system takes a step, then the time progresses until $i \cdot t_1$,

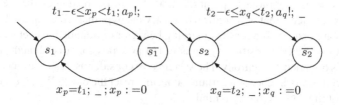

Fig. 5. CTSs for different rates

and the system takes a step that resets the local clock. A computation of Φ_2 is similar, with t_2 replacing t_1.

Let P and Q be two TSs with scheduling variables a_p and a_q as above. Consider the composition of P and Q with Φ_1 and Φ_2. It follows that P moves every t_1 time units and Q every t_2 time units[6]. Every t_1 time units P performs an action, and every t_2 time units Q performs an action. At time t such that both t_1 and t_2 divide t, both P and Q make moves, however, the order between P and Q is not determined. We can show that under appropriate projection, the sequence of actions taken by the composition of the four systems, is equivalent to the sequence of actions taken by the bounded-asynchronous composition of P and Q with rates t_1 and t_2, respectively.

We note that the CTSs have their resets set at exact time points, suggesting that a composition of such systems requires a central clock. We can still maintain 'bounded-asynchronous' behavior if the reset occurs concurrently with the system, however, maintaining ϵ small enough and restricting the number of steps made by the system. For example, if ϵ is $1/100$, then regardless of the exact behavior, the first 98 macro-steps still respect bounded asynchrony. It follows, that unsynchronized local clocks augmented by frequent enough synchronizations would lead to the exact same behavior. It is an interesting question whether similar ideas can be used for the abstraction of real time and probabilistic systems.

Acknowledgments

We thank Alex Hajnal for fruitful discussions, and Marc Schaub for comments on an earlier draft of the manuscript.

References

1. http://mtc.epfl.ch/~piterman/bio
2. Alur, R., Belta, C., Ivancic, F., Kumar, V., Mintz, M., Pappas, G., Rubin, H., Schug, J.: Hybrid modeling and simulation of biomolecular networks. In: Di Benedetto, M.D., Sangiovanni-Vincentelli, A.L. (eds.) HSCC 2001. LNCS, vol. 2034, pp. 19–32. Springer, Heidelberg (2001)
3. Barjis, J., Barjis, I.: Formalization of the protein production by means of petri nets. In: Proceedings International Conference on Information Intelligence Systems, pp. 4–9. IEEE, Los Alamitos (1999)
4. Bernot, G., Comet, J.P., Richard, A., Guespin, J.: Application of formal methods to biological regulatory networks: extending thomas asynchronous logical approach with temporal logic. Journal of Theoretical Biology 229(3), 339–347 (2004)
5. Calzone, L., Chabrier-Rivier, N., Fages, F., Soliman, S.: Machine learning biochemical networks from temporal logic properties. Transactions on Computational Systems Biology 4, 68–94 (2006)
6. Clarke, E.C., Grumberg, O., Peled, D.: Model Checking. MIT Press, Cambridge (1999)

---

[6] For every ϵ and for every values t_1 and t_2 there are some integers i_1 and i_2 such that $[i_1 \cdot t_1 - \epsilon, i_1 \cdot t_1]$ intersects $[i_2 \cdot t_2 - \epsilon, i_2 \cdot t_2]$. As we are interested only in the sequence of actions taken by p_1 and p_2, restricting t_1 and t_2 to range over integer values seems reasonable.

7. Dill, D., Knapp, M.A., Gage, P., Talcott, C., Laderoute, K., Lincoln, P.: The pathalyzer: a tool for analysis of signal transduction pathways. In: Proceedings of the First Annual Recomb Satellite Workshop on Systems Biology (2005)
8. Efroni, S., Harel, D., Cohen, I.R.: Toward rigorous comprehension of biological complexity: modeling, execution, and visualization of thymic T-cell maturation. Genome Res. 13(11), 2485–2497 (2003)
9. Efroni, S., Harel, D., Cohen, I.R.: Emergent dynamics of thymocyte development and lineage determination. PLoS Computational Biology 3(1), 127–136 (2007)
10. Fisher, J., Henzinger, T.A.: Executable cell biology. Nature Biotechnology 25(11), 1239–1249 (2007)
11. Fisher, J., Piterman, N., Hajnal, A., Henzinger, T.A.: Predictive modeling of signalling crosstalk during C. elegans vulval development. PLoS Computational Biology 3(5), 92 (2007)
12. Fisher, J., Piterman, N., Hubbard, E.J., Stern, M.J., Harel, D.: Computational insights into *Caenorhabditis elegans* vulval development. Proc. Natl. Acad. Sci. U S A 102(6), 1951–1956 (2005)
13. Ghosh, R., Tomlin, C.: Lateral inhibition through delta-notch signaling: A piecewise affine hybrid model. In: Di Benedetto, M.D., Sangiovanni-Vincentelli, A.L. (eds.) HSCC 2001. LNCS, vol. 2034, pp. 232–246. Springer, Heidelberg (2001)
14. Gillespie, D.T.: Exact stochastic simulation of coupled chemical reactions. J. of Phys. Chemistry 81(25), 2340–2361 (1977)
15. Heath, J., Kwiatkowska, M., Norman, G., Parker, D., Tymchyshyn, O.: Probabilistic model checking of complex biological pathways. Theoretical Computer Science (2008); Special issue on Converging Sciences: Informatics and Biology (to appear)
16. Henzinger, T.A., Horowitz, B., Kirsch, C.M.: Giotto: A time-triggered language for embedded programming. In: Henzinger, T.A., Kirsch, C.M. (eds.) EMSOFT 2001. LNCS, vol. 2211, pp. 166–184. Springer, Heidelberg (2001)
17. Holzmann, G.J.: The model checker SPIN. IEEE Trans. on Software Engineering 23(5), 279–295 (1997) Special issue on Formal Methods in Software Practice
18. Kam, N., Harel, D., Cohen, I.R.: The immune system as a reactive system: Modeling T-cell activation with statecharts. In: IEEE Symposium of Visual Languages and Formal Methods, pp. 15–22 (2003)
19. Kam, N., Harel, D., Kugler, H., Marelly, R., Pnueli, A., Hubbard, E.J.A., Stern, M.J.: Formal modeling of C. elegans development: A scenario-based approach. In: Priami, C. (ed.) CMSB 2003. LNCS, vol. 2602, pp. 4–20. Springer, Heidelberg (2003)
20. Kauffman, S.A.: Metabolic stability and epigenesis in randomly constructed genetic nets. Journal of Theoretical Biology 22(3), 437–467 (1969)
21. Kesten, Y., Pnueli, A.: Control and data abstractions: The cornerstones of practical formal verification. Software Tools for Technology Transfer 2(1), 328–342 (2000)
22. Priami, C., Regev, A., Shapiro, E.Y., Silverman, W.: Application of a stochastic name-passing calculus to representation and simulation of molecular processes. Information Processing Letters 80(1), 25–31 (2001)
23. Sadot, A., Fisher, J., Barak, D., Admanit, Y., Stern, M.J., Hubbard, E.J., Harel, D.: Towards verified biological models. IEEE Transactions in Computational Biology and Bioinformatics (to appear, 2007)

Computational Probability for Systems Biology

Werner Sandmann[1] and Verena Wolf[2]

[1] University of Bamberg, Germany
werner.sandmann@uni-bamberg.de
[2] EPFL, Switzerland
verena.wolf@epfl.ch

Abstract. Stochastic models of biological networks properly take the randomness of molecular dynamics in living cells into account. Numerical solution approaches inspired by computational methods from applied probability can efficiently yield accurate results and have significant advantages compared to stochastic simulation. Examples for the success of non-simulative numerical analysis techniques in systems biology confirm the enormous potential.

Keywords: Biological Networks, Stochastic Modeling, Markov Chains, Structured Model Representation, Numerical Analysis.

1 Introduction

In order to gain a system-level understanding of intra- and intercellular dynamics, mathematical modeling and efficient computational analysis of biological networks, also referred to as pathways, is a primary scope of systems biology. Just like artificial or technical systems, living systems consist of various mutually related interacting components and are enormously complex. However, the achievements in genomics, proteomics, cell and molecular biology today provide a wealth of data that not only yields the necessary information for suitable modeling but also facilitates the choice of an appropriate abstraction level. From this data, components forming an "isolated" subsystem can be identified leading to an indispensable and necessary reduction of complexity.

As biological systems are constituted by coupled chemical reactions on the molecular level, chemical reactions are essential for all modeling approaches. The fundamental rules are given by stoichiometric equations defining which molecular species may react in order to result in a certain product and how many molecules are involved in the reaction. Quantitative timing aspects are specified by reaction rates assigned to each reaction. Models may be state-continuous or state-discrete and their dynamical behavior may be deterministic or stochastic which is reflected by the most popular instances of deterministic and stochastic reaction kinetics, respectively. In both cases, it is assumed that the system is well stirred and thermally equilibrated, meaning that a well stirred mixture of molecules inside some fixed volume interact at constant temperature. That is, the system is spatially homogeneous such that the concentration or the number of molecules do not depend on positions in space.

J. Fisher (Ed.): FMSB 2008, LNBI 5054, pp. 33–47, 2008.

In deterministic reaction kinetics based on the generalized law of mass actions, the system state at any time is given by the concentrations (measured in mol per liter) of each molecular species. Expressing the system dynamics in terms of deterministic rate equations yields a system of ordinary differential equations (ODEs) for the concentrations. This approach assumes continuous, deterministic changes in concentrations of molecular species and can be suitable for reaction networks involving large populations as, for instance, in metabolic pathways but it neither properly models the discreteness of molecular quantities nor the inherent randomness in chemical reactions. In particular, for gene expression and signal transduction processes it has been extensively demonstrated that stochastic noise plays a major role and should be taken into account. For example, consider a simple bistable genetic toggle switch constructed from repressible promoters. In reality, the system can end up in two different stable states and for most initial configurations both states are possible, that is they have a positive probability that should not be neglected. But the solution of any deterministic reaction rate model yields that in the time limit the system is in exactly one of these two states ignoring the other one completely. Many more examples and studies elucidate the stochastic nature of biological systems such that it is nowadays evident and stochastic models are well established in systems biology [1,5,17,18,30,33,34,38,44,46,47,49].

In the stochastic approach that we focus on, the system state at any time is given by the number of molecules of each species and the system is modeled by a continuous-time Markov chain (CTMC). The system dynamics are described by a system of ODEs called the chemical master equation (CME). Stochastic simulation is in widespread use for analyzing stochastic models of biological networks. It can be applied to arbitrarily large models but it also has a couple of major drawbacks. Stochastic simulation is computationally expensive and can only provide statistical estimates. Rather than directly solving the CME, trajectories (sample paths) of the CTMC are generated. Stochastically exact trajectory generation, often referred to as the Gillespie algorithm in the context of chemical reactions, is exceedingly slow. Even with approximate methods for accelerated trajectory generation a large number of trajectories is required in order to obtain reliable and meaningful results with acceptable statistical accuracy. Hence, if efficient numerical (non-simulative) solutions are possible they should be clearly preferred as they do not suffer from statistical inaccuracies. Standard techniques for numerically solving systems of ODEs become infeasible in case of very large state spaces but it is intuitively clear that exploiting the stochastic interpretation and the specific structure of CTMCs might be helpful.

Computational probability is concerned with efficient methods that are specifically designed for stochastic models. In particular, various sophisticated techniques for the numerical solution of Markov chains have been developed [24,25,45]. However, they are not yet prevalent for solving stochastic models of coupled chemical reactions. In this paper, we demonstrate that they can be successfully used for solving stochastic models of biological systems.

Based on previous work, we present two examples where techniques inspired by methods originally developed for computer systems performance evaluation have been applied to chemically reacting systems. The first example illustrates the model reduction approach of [11]. As the second example, we describe a structured representation of a genetic regulatory network as another promising application of computational probability methods to systems biology, which has been recently proposed in [50].

The remainder of this paper is organized as follows. In Section 2 we introduce stochastic chemical reaction kinetics, describe the corresponding Markovian models and uniformization as a numerical solution approach. The reduction technique for stiff models and the structured representation of a gene feedback loop are presented in Sections 3 and 4, respectively. Finally, Section 5 concludes the paper and outlines directions of further research.

2   Stochastic Chemical Reaction Kinetics

Stochastic approaches to chemical reaction kinetics via Markovian models can be traced back to the study of autocatalytic reactions in the 1940s [16]. In the 1950s, [43] considered chain reactions and some types of coupled reactions, and [2] provided a large body of theory resulting in a series of papers on topics covering sequences of unimolecular and bimolecular reactions, reaction rate constants, and several applications. These early works, though using a different terminology, already implicitly included the CME. Detailed reviews and many more historical references can be found in [4,35,29]. It was also recognized quite early that in the thermodynamic limit, when the number of molecules and the volume approach infinity but the concentrations remain finite, the Markovian and the deterministic approach are equivalent [31,37]. In the 1970s, [21,22] provided a physical justification of Markovian models of coupled chemical reactions, which was later rigorously derived in [23] yielding that it is evidently in accordance with the theory of thermodynamics.

2.1   Model Description

We consider $N \in \mathbb{N}$ molecular species $S_1, S_2, \ldots, S_N$ and $M \in \mathbb{N}$ reaction types $R_1, R_2, \ldots, R_M$. Each reaction R_m, $1 \leq m \leq M$ is defined by a reaction equation

$$\ell_1^{(m)} S_1 + \ell_2^{(m)} S_2 + \ldots + \ell_N^{(m)} S_N \xrightarrow{c_m} h_1^{(m)} S_1 + h_2^{(m)} S_2 + \ldots + h_N^{(m)} S_N. \quad (1)$$

The stoichiometric coefficients $\ell_1^{(m)}, \ell_2^{(m)}, \ldots, \ell_N^{(m)} \in \mathbb{N}$ describe for each species how many molecules are consumed if a reaction of type R_m occurs. Similarly, the stoichiometric coefficients $h_1^{(m)}, \ldots, h_N^{(m)} \in \mathbb{N}$ describe how many molecules of each species are produced by R_m. The reaction rate constant $c_m \in \mathbb{R}_{>0}$ determines the "speed" of R_m as explained below. Note that the population of species S_i is unaffected by R_m if $\ell_i^{(m)} - h_i^{(m)} = 0$.

The stochastic process that represents the temporal evolution of the species'
populations is given by a family $\big(X(t)\big)_{t \geq 0}$ of random vectors

$$X(t) = \big(X_1(t), X_2(t), \ldots, X_N(t)\big)$$

taking values in a discrete set $\mathcal{X} \subset \mathbb{N}^N$. The random variable $X_i(t), 1 \leq i \leq N$
describes the number of molecules of species S_i at time instant t. We fix the
initial conditions of the process by defining that $P\big(X(0) = \boldsymbol{x}_0\big) = 1$ for an initial
population vector $\boldsymbol{x}_0 \in \mathcal{X}$. The *transient state probability* that at time $t \geq 0$ the
system is in state $\boldsymbol{x} = (x_1, x_2, \ldots, x_N)$, given $X(0) = \boldsymbol{x}_0$, is denoted by

$$p^{(t)}(\boldsymbol{x}) = P\big(X(t) = \boldsymbol{x} \mid X(0) = \boldsymbol{x}_0\big). \tag{2}$$

State changes are triggered by chemical reactions. For an infinitesimal time interval $[t, t + dt)$

$$P\big(R_m \text{ occurs in } [t, t + dt) \mid X(t) = \boldsymbol{x}\big) = \alpha_m(\boldsymbol{x}) \cdot dt \tag{3}$$

where $\alpha_m : \mathcal{X} \to \mathbb{R}_{\geq 0}$ is called the *propensity function* of R_m. This probability is
proportional to the number of distinct combinations of R_m's reactants. Hence,
$\alpha_m(\boldsymbol{x})$ computes as

$$\alpha_m(\boldsymbol{x}) = c_m \cdot \prod_{j=1}^{N} \binom{x_j}{\ell_j^{(m)}}. \tag{4}$$

The probability in (3) only depends on the length of the time interval which
means that the propensity functions are time-independent. Besides, the next
state in the system's time evolution only depends on the current state, and neither on the specific time nor on the history of reactions that led to the current
state. Hence, the system is in fact modeled as a (time-homogeneous, conserva-
tive) CTMC $(X(t))_{t \geq 0}$ with N-dimensional state space $\mathcal{X} \subseteq \mathbb{N}^N$. This gives rise
to a state-transition graph representation in which outgoing transitions of $\boldsymbol{x}$ are
labeled by *transition rates* $\alpha_m(\boldsymbol{x})$. The successor state according to a transition
from $\boldsymbol{x}$ triggered by a reaction of type R_m is state $\boldsymbol{x} + \boldsymbol{v}_m$ where vector $\boldsymbol{v}_m$
equals the m-th row of the *stoichiometric matrix* $V \in \mathbb{Z}^{M \times N}$, defined as follows:
if R_m is given by (1) the m-th row of V is such that the i-th entry contains the
difference of the number of S_i molecules after a reaction of type R_m occured,
i.e. $v_{mi} = h_i^{(m)} - \ell_i^{(m)}$.

2.2   Chemical Master Equation and Kolmogorov Differential Equations

The system dynamics in terms of the state probabilities' time derivatives are
given by the *chemical master equation* (CME)

$$\frac{\partial p^{(t)}(\boldsymbol{x})}{\partial t} = \sum_{m=1}^{M} \big(\alpha_m(\boldsymbol{x} - \boldsymbol{v}_m)p^{(t)}(\boldsymbol{x} - \boldsymbol{v}_m) - \alpha_m(\boldsymbol{x})p^{(t)}(\boldsymbol{x})\big). \tag{5}$$

Terminology and notation as introduced here and commonly used in systems biology as well as in chemistry and physics is quite different from that in mathematics, computer science, and engineering. In particular, it is a purely functional specification as opposed to an algebraic matrix specification. Consequently, expressions governing the system dynamics usually adhere to one of these specifications. At a first glance they may appear to be rather different but of course they are equivalent. More specifically, as we will see below the chemical master equation is one way to write the Kolmogorov differential equations.

In order to recognize the equivalence of the functional specification and a matrix specification note that the multidimensional discrete state space can be mapped to the set $\mathbb{N}$ of nonnegative integers such that each state $x \in \mathcal{X}$ is uniquely assigned to an integer $i \in \mathbb{N}$. The probability that a transition from state $i \in \mathbb{N}$ to state $j \in \mathbb{N}$ occurs within a time interval of length $h \geq 0$ is denoted by $p_{ij}(h)$, and correspondingly $P(h) = (p_{ij}(h))_{i,j \in \mathbb{N}}$ is a stochastic matrix, where $P(0)$ equals the unit matrix I, since no state transitions occur within a time interval of length zero. It is well known (cf. [7,29]) that a CTMC is uniquely defined by an initial probability distribution and a transition rate matrix, also referred to as *infinitesimal generator matrix*, $Q = (q_{ij})_{i,j \in \mathbb{N}}$ consisting of transition rates q_{ij} where Q is the derivative at 0 of the matrix function $h \mapsto P(h)$. The relation of each $P(h)$ to Q is given by $P(h) = \exp(Qh)$. In that way Q generates the transition probability matrices by a matrix exponential function which is basically defined as an infinite power series. Hence, all information on transition probabilities is covered by the single matrix Q. In terms of Q the *Kolmogorov global differential equations* can be expressed by

$$\frac{\partial \boldsymbol{p}^{(t)}}{\partial t} = \boldsymbol{p}^{(t)} Q \tag{6}$$

where $\boldsymbol{p}^{(t)} = (p_1^{(t)}, p_2^{(t)}, \ldots)$ denotes the vector of the transient state probabilities corresponding to Equation (2). Explicitly writing each row of Equation (6) yields

$$\frac{\partial p_i^{(t)}}{\partial t} = \sum_{j:j \neq i} p_j^{(t)} q_{ji} - \sum_{j:j \neq i} p_i^{(t)} q_{ij} = \sum_{j:j \neq i} \left(p_j^{(t)} q_{ji} - p_i^{(t)} q_{ij} \right) \tag{7}$$

from which the equivalence to the CME is now easily seen by interpreting $i \in \mathbb{N}$ as the number assigned to state $x \in \mathcal{X}$, i.e. $p_i^{(t)} = p^{(t)}(x)$, $q_{ij} = \alpha_m(x)$ if j is the number assigned to state $x + v_m$, and $q_{ji} = \alpha_m(x - v_m)$ if j is the number assigned to state $x - v_m$.

2.3 Uniformization

A general solution of Equation (6) is [15,25,45]

$$\boldsymbol{p}^{(t)} = \boldsymbol{p}^{(0)} \cdot P(t) = \boldsymbol{p}^{(0)} \cdot \exp(Qt) = \boldsymbol{p}^{(0)} \cdot \sum_{k=0}^{\infty} \frac{(Qt)^k}{k!}. \tag{8}$$

The infinite sum includes matrix exponentials and its computation is numerically unstable as Q contains strictly positive and negative entries leading to severe round-off errors [36]. As already mentioned in the introduction, different methods exist to compute the vector $p^{(t)}$. The most common one is the uniformization method which is based on the Taylor series expansion of the matrix exponential. It goes back to Jensen [28] and is thus sometimes referred to as Jensen's method. It is also known as *randomization* or *discrete-time conversion* and has been applied to computing transient as well as steady-state solutions for Markov chains, see for example [15,26,27,45]. The basic idea is to define an associated discrete-time Markov chain (DTMC) that behaves equivalently to the CTMC $(X(t))_{t\geq 0}$ in the sense that it is stochastically identical. More specifically, the CTMC is represented as a DTMC where the times are implicitly driven by a Poisson process. We briefly present it as it will be used later on.

Define a *uniformization rate* λ such that

$$\lambda \geq \sup_{1\leq i\leq n} \sum_{j:j\neq i} q_{ij}$$

and construct the stochastic matrix

$$P = I + \frac{1}{\lambda}Q$$

which is the transition matrix of the associated DTMC. The matrix P^k contains the k-step transition probabilities. Thus, $w^{(k)} := p^{(0)}P^k$ is the vector of the state probabilities after k steps in the DTMC. The probability of k steps within the time interval $[0, t)$ has a Poisson distribution with parameter λt, i.e.

$$P\big(k \text{ steps until time } t\big) = e^{-\lambda t}\frac{(\lambda t)^k}{k!}. \tag{9}$$

Now, the solution of the transient state probabilities in Equation (8) can be rewritten (cf. [15,25,45]) as

$$p^{(t)} = p^{(0)} \cdot \sum_{k=0}^{\infty} e^{-\lambda t}\frac{(\lambda t)^k}{k!}P^k = \sum_{k=0}^{\infty} e^{-\lambda t}\frac{(\lambda t)^k}{k!} \cdot w^{(k)}. \tag{10}$$

Equation (10) has nice properties compared to (8). There are no negative summands involved as P is a stochastic matrix and $\lambda > 0$. Moreover, $w^{(k)}$ can be computed recursively by

$$w^{(0)} = p^{(0)}, \quad w^{(j)} = w^{(j-1)} \cdot P, \quad j \in \{1, 2, \ldots\}.$$

If P is sparse, $w^{(k)}$ can be calculated efficiently even if the size n of the state space is large. Note that it is possible to obtain $p^{(t)}$ via Equation (10) for several values of t simultaneously where the $w^{(k)}$ are only computed once. According to [20] left and right summation bounds L and R can be obtained such that the truncation error

$$p^{(t)} - \hat{p}^{(t)} := \left(\sum_{k=0}^{\infty} e^{-\lambda t}\frac{(\lambda t)^k}{k!}w^{(k)}\right) - \left(\sum_{k=L}^{R} e^{-\lambda t}\frac{(\lambda t)^k}{k!}w^{(k)}\right) < \epsilon$$

can be a priori bounded by a predefined $\epsilon > 0$. Thus, $p^{(t)}$ can be approximated arbitrarily accurate as long as the required number of summands is not extremely large, and diverse properties of interest can be obtained from $p^{(t)}$. In many cases, the uniformization approach is superior to other numerical analysis techniques.

In particular, uniformization can be applied to networks of coupled chemical reactions. For example, [41,42] formulated it in terms of the functional representation and also presented a discrete-time version of the chemical master equation. However, a corresponding stochastic simulation algorithm was presented rather than applying uniformization for non-simulative numerical analysis.

3 Numerical Aggregation for a Stiff Enzymatic Reaction

For both deterministic and stochastic models, the mathematical property of *stiffness* renders system analysis difficult and often even impossible with traditional methods. Stiffness arises whenever the components of the underlying system act on time scales that differ by several orders of magnitude which is typically the case for biological systems. Standard numerical methods as well as stochastic simulation perform extremely ineffecient in the presence of stiffness and advanced solution approaches are required.

We consider the enzyme-catalyzed substrate conversion

$$
\begin{aligned}
R_1 : E + S &\xrightarrow{c_1} C \\
R_2 : \quad C &\xrightarrow{c_2} E + S \\
R_3 : \quad C &\xrightarrow{c_3} E + P
\end{aligned}
\tag{11}
$$

of a substrate S into a product P via an enzyme-substrate complex C, catalyzed (accelerated) by an enzyme E. Usually, the initial number of substrate molecules is large compared to the small enzyme population. Typical measures of interest are the mean and the variance of the time that is needed until all substrate molecules are transformed into product molecules. If reactions of type R_2 are much faster than those of type R_3, the reaction set becomes stiff. Here, this is the case for $c_2 \gg c_3$. In general, stiffness can not always be identified by simply inspecting the rate constants c_m but one has to inspect the transition rates, that is the state dependent propensity functions α_m.

In [11] a numerical aggregation algorithm (NAA) for the enzyme-catalyzed substrate conversion is proposed. The idea is that if the stiffness condition is valid an appropriately modified adaptation of the aggregation technique in [6] yields an accurate approximation of the desired measures. The method essentially consists of two steps:

1. The state space of the CTMC is partitioned into subsets of states, called aggregates, such that within an aggregate the states are connected via transitions of reaction types R_1 and R_2 only. These subsets, considered in isolation, form the "fast" part of the model and are "almost in steady-state". The relative percentage of the time spent in each state is computed in the limit as

time approaches infinity. Based on these probabilities an aggregated model is constructed in which each aggregate forms a "macro state". The transition rates between the macro states are the weighted sums of the individual cumulative transition rates in the original model.

2. As each subset forms a macro state, the aggregated model is much smaller than the original one. Moreover, stiffness is eliminated because all reactions in the aggregated model occur at the same time scale. Hence, uniformization can be applied to the aggregated model resulting in an approximation for the original one. Since for the enzyme-catalyzed substrate conversion only transitions via R_3 remain, the model has the structure of a simple birth process. Techniques that exploit this simple structure yield accurate results very efficiently.

Let $x_0(E)$ and $x_0(S)$ be the initial population size of the species E and S, respectively, and $n = x_0(S) + 1$. Aggregate A_k, $1 \le k \le n$ is then given by

$$A_k = \{(x_E, x_S, x_C, x_P) \mid x_P = k - 1, x_E + x_C = x_0(E), x_S + x_C + x_P = x_0(S)\}.$$

Thus, transitions between different aggregates are triggered by reactions of type R_3 whereas each transition within an aggregate corresponds to a reaction of type R_1 or type R_2. Let $Q^{(k)}$ be the submatrix of Q which contains the entries of the elements of A_k, that is $Q^{(k)}$ is the generator matrix of the CTMC with state space A_k. The *steady-state distribution* of A_k is obtained as the unique solution of the linear system

$$\pi^{(k)}Q^{(k)} = 0, \quad \sum_{i \in A_k} \pi_i^{(k)} = 1.$$

The aggregated CTMC has state space $\{A_1, \ldots, A_n\}$, that is the states are subsets of the state space of the original CTMC. The transition rate between macro states A_k and $A_{k'}$, $k \ne k'$ is defined as

$$\hat{q}_{kk'} := \sum_{i \in A_k} \pi_i^{(k)} \sum_{j \in A_{k'}} q_{ij}.$$

By setting

$$\hat{q}_{ii} := - \sum_{j=1, j \ne i}^{n} \hat{q}_{ij} \text{ and } \hat{Q} := (\hat{q}_{ij})_{1 \le i, j \le n}$$

a smaller CTMC is obtained for which transient state probabilities are calculated using uniformization.

Experimental results show that for stiff systems, the NAA is very accurate and fast. For example, comparisons with an approximate stochastic simulation algorithm that was specifically designed for stiff systems, the slow-scale stochastic simulation algorithm [12,13], show that the NAA is at least ten times faster and even up to more than 10^4 times faster in parameter regions where only a small number of enzyme molecules is present.

4   Structured Representation of a Gene Regulatory Network

As outlined in Section 2, besides the functional description in terms of propensities, the underlying CTMC of a biochemical network can be represented by the corresponding generator matrix Q and the initial distribution. Various numerical solution algorithms are based on an explicit matrix description which requires the construction of Q in order to apply these methods. Unfortunately, the size of Q grows exponentially in the number of involved species which means that the analysis suffers from the so-called *state space explosion problem*. The number of non-zero entries in Q might be so large that the computer storage capacity is exceeded.

In this section, we give a structured Kronecker-based representation that describes the underlying CTMC of a biochemical network. The generator matrix is expressed in a modular way such that the state space explosion problem is circumvented. In general, the Kronecker representation can be used to model very large CTMCs whose state space is in the order of billions. The basis is a generalized tensor algebra with a *Kronecker product* operation [48]. In the context of biochemical networks, it is well suited for expressing reaction rates.

The Kronecker product of two matrices $A = (a_{ij}) \in \mathbb{R}^{n_1 \times m_1}$ and $B = (b_{kl}) \in \mathbb{R}^{n_2 \times m_2}$ is defined as $C = A \otimes B$, where C is an $n_1 \times m_1$ block matrix whose (i, j) block is the $n_2 \times m_2$ matrix $a_{ij}B$. For illustration, consider $A \in \mathbb{R}^{2 \times 2}$ and $B \in \mathbb{R}^{3 \times 3}$. Their Kronecker product $C = A \otimes B \in \mathbb{R}^{6 \times 6}$ is given by

$$
C = \begin{pmatrix} a_{11}\,B & a_{12}\,B \\ a_{21}\,B & a_{22}\,B \end{pmatrix} = \left(\begin{array}{ccc|ccc}
a_{11}b_{11} & a_{11}b_{12} & a_{11}b_{13} & a_{12}b_{11} & a_{12}b_{12} & a_{12}b_{13} \\
a_{11}b_{21} & a_{11}b_{22} & a_{11}b_{23} & a_{12}b_{21} & a_{12}b_{22} & a_{12}b_{23} \\
a_{11}b_{31} & a_{11}b_{32} & a_{11}b_{33} & a_{12}b_{31} & a_{12}b_{32} & a_{12}b_{33} \\
\hline
a_{21}b_{11} & a_{21}b_{12} & a_{21}b_{13} & a_{22}b_{11} & a_{22}b_{12} & a_{22}b_{13} \\
a_{21}b_{21} & a_{21}b_{22} & a_{21}b_{23} & a_{22}b_{21} & a_{22}b_{22} & a_{22}b_{23} \\
a_{21}b_{31} & a_{21}b_{32} & a_{21}b_{33} & a_{22}b_{31} & a_{22}b_{32} & a_{22}b_{33}
\end{array} \right).
$$

The Kronecker product is associative and distributive with respect to matrix addition. A variety of numerical algorithms for Kronecker-based representations of CTMCs exist and are implemented in tools like PEPS [3], APNN [9] and SMART [14].

Note that the Kronecker representation can be generated from high-level modeling paradigms for Markov chains. In particular for stochastic automata networks (SANs) [32,39,40], it is well suited. A SAN consists of a number of interacting stochastic automata that are described by local state-transition graphs. An automaton takes local transitions independently of the other automata whereas for synchronizing transitions two or more automata change their state. In the Kronecker representation, this is directly reflected by local matrices that are combined by Kronecker operations. For more information on SANs we refer to [32,39,40]. Here, we focus on the Kronecker representation.

Consider a reaction network with $N \in \mathbb{N}$ molecular species $S_1, S_2, \ldots, S_N$ and $M \in \mathbb{N}$ reaction types $R_1, R_2, \ldots, R_M$. Let $N_i \in \mathbb{N}$ be the maximum number of molecules of species S_i and let reaction type R_m be given by (1). The idea is to represent the generator matrix Q implicitly as a Kronecker product of (smaller) component matrices. Define the vector

$$\boldsymbol{u}^{(m,i)} = \left(\binom{0}{\ell_i^{(m)}}, \binom{1}{\ell_i^{(m)}}, \dots, \binom{N_i}{\ell_i^{(m)}} \right) \in \mathbb{N}^{N_i+1}.$$

whose entries are the binomial coefficients $\binom{j}{\ell_i^{(m)}}$, $0 \le j \le N_i$. These are the factors by which S_i contributes to the calculation of the transition rate, that is the propensity function α_m according to (4). If $\ell_i^{(m)} = 0$, the transition rate of the reaction is independent of the current population of S_i, i.e. the contributed multiplicative factor is one.

Given $\boldsymbol{u}^{(m,i)}$, a matrix $U^{(m,i)}$ that corresponds to reaction type R_m and species S_i is constructed as follows: the vector $\boldsymbol{u}^{(m,i)}$ appears at the v_{mi}-th diagonal (upper diagonal if $v_{mi} > 0$ and lower diagonal otherwise) of $U^{(m,i)}$ if reaction R_m increases (decreases) the population of S_i by v_{mi} (the entry in the stoichiometric matrix V). All remaining entries of $U^{(m,i)}$ are zero. Note that we count the positions of the diagonals such that $\boldsymbol{u}^{(m,i)}$ appears at the main diagonal of $U^{(m,i)}$ if $v_{mi} = 0$.

Consider a two-gene positive feedback loop with $M = 8$ reactions involving $N = 4$ chemical species. With $k, r \in \{1, 2\}, k \ne r$ the reactions are given by[1]

$$
\begin{array}{lll}
R_k^{tcr} & : \text{Prot}_r \xrightarrow{c_k^{tcr}} \text{mRNA}_k + \text{Prot}_r & \text{(transcription of mRNA}_k\text{)} \\
R_r^{tsl} & : \text{mRNA}_r \xrightarrow{c_r^{tsl}} \text{mRNA}_r + \text{Prot}_r & \text{(translation of Prot}_r\text{)} \\
R_r^{mdeg} & : \text{mRNA}_r \xrightarrow{c_r^{mdeg}} \emptyset & \text{(degradation of mRNA}_r\text{)} \\
R_r^{pdeg} & : \text{Prot}_r \xrightarrow{c_r^{pdeg}} \emptyset & \text{(degradation of Prot}_r\text{)}
\end{array}
\tag{12}
$$

This set of reactions describes a small regulatory network controlling the transcription of two genes into mRNA and the translation of the two corresponding types of mRNA into proteins. The transcription of gene 1 depends on the population of promoter Prot_2 (reaction R_1^{tcr}). The Prot_2 molecules, in turn, are translation products of mRNA_2 (R_2^{tsl}) and Prot_1 results from the translation of mRNA_1 (R_1^{tsl}). The Prot_1 molecules act as regulatory proteins for the transcription of gene 2 (R_2^{tcr}). Molecules degrade according to the reactions R_r^{mdeg} and R_r^{pdeg}. For transcription of mRNA_k, that is $R_m = R_k^{tcr}$, $S_i = \text{mRNA}_k$ and $S_j = \text{Prot}_r$ we have the matrices

$$
U^{(m,i)} = \begin{pmatrix} 0 & 1 & & \cdots & 0 \\ 0 & 0 & 1 & \cdots & 0 \\ 0 & 0 & 0 & \ddots & \vdots \\ \vdots & \vdots & \ddots & \ddots & 1 \\ 0 & 0 & \cdots & 0 & 0 \end{pmatrix}, \quad U^{(m,j)} = \begin{pmatrix} 0 & 0 & & \cdots & 0 \\ 0 & 1 & 0 & \cdots & 0 \\ 0 & 0 & 2 & \ddots & \vdots \\ \vdots & \vdots & \ddots & \ddots & 0 \\ 0 & 0 & \cdots & 0 & N_j \end{pmatrix}.
$$

---

[1] The symbol $\emptyset$ on the right-hand indicates that the number of products is zero.

We obtain the same matrices for the translation, i.e. $S_i = \mathsf{Prot}_k$, $R_m = R_k^{tsl}$ and $S_j = \mathsf{mRNA}_k$. Degradation of mRNA_k or Prot_k yields

$$
U^{(m,i)} = \begin{pmatrix} 0 & 0 & & \cdots & 0 \\ 1 & 0 & 0 & \cdots & 0 \\ 0 & 2 & 0 & \ddots & \vdots \\ \vdots & \vdots & \ddots & \ddots & 0 \\ 0 & 0 & \cdots & N_j & 0 \end{pmatrix}.
$$

Note that the vector $\boldsymbol{u}^{(m,i)}$ is truncated if it does not appear on the main diagonal as $U^{(m,i)}$ is always of size $N_i + 1$. We set $D^{(m,i)} = diag(U^{(m,i)}e^T)$ where $\boldsymbol{e}$ is a unit row vector of appropriate size and the operator $diag(\boldsymbol{u})$ constructs a diagonal matrix from the vector $\boldsymbol{u}$, i.e. $\boldsymbol{u}$ appears on the main diagonal. Now, the generator matrix Q of the CTMC can be expressed as

$$
Q = \sum_{m=1}^{M} c_m \left(\bigotimes_{i=1}^{N} U^{(m,i)} - \bigotimes_{i=1}^{N} D^{(m,i)} \right) = \sum_{m=1}^{M} c_m \left(\bigotimes_{i=1}^{N} U^{(m,i)} - D \right) \tag{13}
$$

where $\otimes$ denotes the Kronecker product operation. Note that subtracting the $D^{(m,i)}$ ensures that the row sums of Q are zero. Moreover, their Kronecker product D is also a diagonal matrix. The matrix Q agrees with the generator matrix defined in Section 2 up to the ordering of the states. The number of factors used in the Kronecker representation of a biochemical network grows only linearly in the number of involved species and reactions and is independent of the population size. The sizes of the individual matrices depend on the maximum numbers of molecules of the participating molecular species.

The Kronecker representation of Q can be exploited for analysis if the uniformization method is applied. From (10) we have

$$
\boldsymbol{p}^{(t)} = \sum_{k=0}^{\infty} e^{-\lambda t} \frac{(\lambda t)^k}{k!} \cdot \boldsymbol{w}^{(k)}
$$

where $\boldsymbol{w}^{(0)} = \boldsymbol{p}^{(0)}$ and for $j \in \{1, 2, \ldots\}$

$$
\begin{aligned}
\boldsymbol{w}^{(j)} &= \boldsymbol{w}^{(j-1)} \cdot P \\
&= \boldsymbol{w}^{(j-1)} \cdot (I + \tfrac{1}{\lambda}Q) \\
&= \boldsymbol{w}^{(j-1)} + \tfrac{1}{\lambda}\boldsymbol{w}^{(j-1)} \cdot Q \\
&\stackrel{(13)}{=} \boldsymbol{w}^{(j-1)} + \tfrac{1}{\lambda} \sum_{m=1}^{M} c_m \, \boldsymbol{w}^{(j-1)} \cdot \left(\bigotimes_{i=1}^{N} U^{(m,i)} - D \right).
\end{aligned}
$$

Efficient techniques exist for vector matrix multiplications if the matrix has Kronecker representation, see, for instance, [8,19] and the references therein. Most of them exploit the fact that usually the matrices $U^{(m,i)}$ are sparse which is in particular the case if the system under study is a network of biochemical reactions. Thus, even in the case of a large state space numerical techniques based on the matrix representation Q can be applied. The CTMC described by Q may contain many states unreachable from the initial one. We refer to [10] for an efficient way to determine the subset of reachable states in the potential state space.

As future work, we plan to carry out case studies based on realistic biological examples, such as the virus model of [44], in order to emphasize the feasibility and the practical relevance of the Kronecker approach.

5   Conclusion

The stochastic modeling approach provides significant insights into biochemical pathways. The analysis of the underlying Markov chain via standard ODE solvers or stochastic simulation is computationally demanding and often becomes intractable for complex models. In particular, large state spaces and multiple time scales or stiffness pose grand challenges. We suggest to built on efficient computational probability methods that were originally developed for other application domains, e.g., computer systems performance evaluation.

The numerical aggregation algorithm for the enzyme-catalyzed substrate conversion exploits model reduction techniques such that both largeness and stiffness are removed in the aggregated model. Experimental results show that the algorithm is superior to other approaches with respect to time complexity and accuracy. Further research includes the generalization of the algorithm such that it can be applied to arbitrary networks of biochemical reactions.

The structured representation of the genetic feedback loop via Kronecker algebra provides a modular design process which is adequate for abstraction purposes. The elegant and compact matrix representation keeps track of the network structure and facilitates numerical analysis algorithms that overcome the state space explosion problem with efficient storage mechanisms. Future work in this area includes more case studies for large biological networks as well as the use of already existing tools for the analysis of Kronecker-based representations.

Hence, computational probability is a promising approach to tackle some of the major problems in the analysis of stochastic models in systems biology. First steps were already successful and demonstrated the great potential. Ongoing research projects are likely to contribute substantial progress in advancing systems biology.

References

1. Arkin, A., Ross, J., McAdams, H.H.: Stochastic kinetic analysis of developmental pathway bifurcation in phage λ-infected escherichia coli cells. Genetics 149, 1633–1648 (1998)
2. Bartholomay, A.F.: A Stochastic Approach to Chemical Reaction Kinetics. Phd thesis, Harvard University (1957)

3. Benoit, A., Fernandes, P., Plateau, B., Stewart, W.J.: The PEPS software tool. In: Kemper, P., Sanders, W.H. (eds.) TOOLS 2003. LNCS, vol. 2794, pp. 98–115. Springer, Heidelberg (2003)
4. Bharucha-Reid, A.T.: Elements of the Theory of Markov Processes and Their Applications. McGraw-Hill, New York (1960)
5. Blake, W.J., Kaern, M., Cantor, C.R., Collins, J.J.: Noise in eukaryotic gene expression. Nature 422, 633–637 (2003)
6. Bobbio, A., Trivedi, K.S.: An aggregation technique for the transient analysis of stiff Markov chains. IEEE Transactions on Computers C-35(9), 803–814 (1986)
7. Bremaud, P.: Markov Chains. Springer, Heidelberg (1998)
8. Buchholz, P., Ciardo, G., Donatelli, S., Kemper, P.: Complexity of memory-efficient Kronecker operations with applications to the solution of Markov models. Journal on Computing 12(3), 203–222 (2000)
9. Buchholz, P., Kemper, P.: A toolbox for the analysis of discrete event dynamic systems. In: Halbwachs, N., Peled, D.A. (eds.) CAV 1999. LNCS, vol. 1633, pp. 483–486. Springer, Heidelberg (1999)
10. Buchholz, P., Kemper, P.: Efficient computation and representation of large reachability sets for composed automata. Discrete Event Dynamic Systems 12(3), 265–286 (2002)
11. Busch, H., Sandmann, W., Wolf, V.: A numerical aggregation algorithm for the enzyme-catalyzed substrate conversion. In: Priami, C. (ed.) CMSB 2006. LNCS (LNBI), vol. 4210, pp. 298–311. Springer, Heidelberg (2006)
12. Cao, Y., Gillespie, D.T., Petzold, L.R.: Accelerated stochastic simulation of the stiff enzyme-substrate reaction. Journal of Chemical Physics 123(14), 144917 (2005)
13. Cao, Y., Gillespie, D.T., Petzold, L.R.: The slow-scale stochastic simulation algorithm. Journal of Chemical Physics 122, 14116 (2005)
14. Ciardo, G., Miner, A.: SMART: The stochastic model checking analyzer for reliability and timing. In: Proceedings of the 1st International Conference on Quantitative Evaluation of Systems, pp. 338–339 (2004)
15. de Souza e Silva, E., Gail, R.: Transient solutions for Markov chains. In: Grassmann, W.K. (ed.) Computational Probability, ch. 3, pp. 43–79. Kluwer Academic Publishers, Dordrecht (2000)
16. Delbrück, M.: Statistical fluctuations in autocatalytic reactions. Journal of Chemical Physics 8, 120–124 (1940)
17. Elowitz, M.B., Levine, M.J., Siggia, E.D., Swain, P.S.: Stochastic gene expression in a single cell. Science 297, 1183–1186 (2002)
18. Fedoroff, N., Fontana, W.: Small numbers of big molecules. Science 297, 1129–1131 (2002)
19. Fernandes, P., Plateau, B., Stewart, W.J.: Efficient descriptor-vector multiplications in stochastic automata networks. Journal of the ACM 45(3), 381–414 (1998)
20. Fox, B.L., Glynn, P.W.: Computing Poisson probabilities. Communications of the ACM 31(4), 440–445 (1988)
21. Gillespie, D.T.: A general method for numerically simulating the time evolution of coupled chemical reactions. Journal of Computational Physics 22, 403–434 (1976)
22. Gillespie, D.T.: Exact stochastic simulation of coupled chemical reactions. Journal of Physical Chemistry 81(25), 2340–2361 (1977)
23. Gillespie, D.T.: A rigorous derivation of the chemical master equation. Physica A 188, 404–425 (1992)
24. Grassmann, W.K.: Computational methods in probability theory. In: Heyman, D.P., Sobel, M.J. (eds.) Stochastic Models. Handbooks in Operations Research and Management Science, vol. 2, ch. 5, pp. 199–254. Elsevier, Amsterdam (1990)

25. Grassmann, W.K. (ed.): Computational Probability. Kluwer Academic Publishers, Dordrecht (2000)
26. Gross, D., Miller, D.: The randomization technique as a modeling tool and solution procedure for transient Markov processes. Operations Research 32(2), 926–944 (1984)
27. Hordijk, A., Iglehart, D.L., Schassberger, R.: Discrete time methods for simulating continuous time Markov chains. Advances in Applied Probability 8, 772–788 (1976)
28. Jensen, A.: Markoff chains as an aid in the study of Markoff processes. Skandinavisk Aktuarietidskrift 36, 87–91 (1953)
29. van Kampen, N.G.: Stochastic Processes in Physics and Chemistry, 3rd edn. Elsevier, Amsterdam (2007)
30. Kierzek, A., Zaim, J., Zielenkiewicz, P.: The effect of transcription and translation initiation frequencies on the stochastic fluctuations in prokaryotic gene expression. Journal of Biological Chemistry 276(11), 8165–8172 (2001)
31. Kurtz, T.G.: The relationship between stochastic and deterministic models for chemical reactions. Journal of Chemical Physics 57(7), 2976–2978 (1972)
32. Langville, A.N., Stewart, W.J.: The Kronecker product and stochastic automata networks. Journal of Computational and Applied Mathematics 167(2), 429–447 (2004)
33. McAdams, H.H., Arkin, A.: Stochastic mechanisms in gene expression. Proceedings of the National Academy of Science (PNAS) USA 94, 814–819 (1997)
34. McAdams, H.H., Arkin, A.: It's a noisy business? Trends in Genetics 15(2), 65–69 (1999)
35. McQuarrie, D.A.: Stochastic approach to chemical kinetics. Journal of Applied Probability 4, 413–478 (1967)
36. Moler, C.B., Van Loan, C.F.: Nineteen dubious ways to compute the exponential of a matrix. SIAM Review 20(4), 801–836 (1978)
37. Oppenheim, I., Shuler, K.E., Weiss, G.H.: Stochastic and deterministic formulation of chemical rate equations. Journal of Chemical Physics 50(1), 460–466 (1969)
38. Paulsson, J.: Summing up the noise in gene networks. Nature 427(6973), 415–418 (2004)
39. Plateau, B.: On the stochastic structure of parallelism and synchronization models for distributed algorithms. In: Proceedings of the Sigmetrics Conference on Measurement and Modeling of Computer Systems, pp. 147–154 (1985)
40. Plateau, B., Stewart, W.J.: Stochastic automata networks. In: Grassmann, W.K. (ed.) Computational Probability, ch. 5, pp. 113–152. Kluwer Academic Publishers, Dordrecht (2000)
41. Sandmann, W.: Stochastic simulation of biochemical systems via discrete-time conversion. In: Proceedings of the 2nd Conference on Foundations of Systems Biology in Engineering, pp. 267–272. Fraunhofer IRB Verlag (2007)
42. Sandmann, W.: Discrete-time stochastic modeling and simulation of biochemical networks. Computational Biology and Chemistry (to appear, 2008)
43. Singer, K.: Application of the theory of stochastic processes to the study of irreproducible chemical reactions and nucleation processes. Journal of the Royal Statistical Society, Series B 15(1), 92–106 (1953)
44. Srivastava, R., You, L., Summers, J., Yin, J.: Stochastic vs. deterministic modeling of intracellular viral kinetics. Journal of Theoretical Biology 218, 309–321 (2002)
45. Stewart, W.J.: Introduction to the Numerical Solution of Markov Chains. Princeton University Press, Princeton (1995)

46. Swain, P.S., Elowitz, M.B., Siggia, E.D.: Intrinsic and extrinsic contributions to stochasticity in gene expression. Proceedings of the National Academy of Science (PNAS) USA 99(20), 12795–12800 (2002)
47. Turner, T.E., Schnell, S., Burrage, K.: Stochastic approaches for modelling in vivo reactions. Computational Biology and Chemistry 28, 165–178 (2004)
48. Van Loan, C.F.: The ubiquitous Kronecker product. Journal of Computational and Applied Mathematics 123, 85–100 (2000)
49. Wilkinson, D.J.: Stochastic Modelling for Systems Biology. Chapman & Hall, Boca Raton (2006)
50. Wolf, V.: Modelling of biochemical reactions by stochastic automata networks. Electronic Notes in Theoretical Computer Science 171(2), 197–208 (2007)

Design Issues for Qualitative Modelling of Biological Cells with Petri Nets

Elzbieta Krepska, Nicola Bonzanni,
Anton Feenstra, Wan Fokkink, Thilo Kielmann,
Henri Bal, and Jaap Heringa

Department of Computer Science, Vrije Universiteit
De Boelelaan 1083, 1081 HV Amsterdam, The Netherlands
ekr@cs.vu.nl, {bonzanni,feenstra,wanf}@few.vu.nl,
{kielmann,bal,heringa}@cs.vu.nl

Abstract. Petri nets are a widely used formalism to qualitatively model concurrent systems such as a biological cell. We present techniques for modelling biological processes as Petri nets for further analyses and in-silico experiments. Instead of extending the formalism with ,,colours" or rates, as is most often done, we focus on preserving the simplicity of the formalism and developing an execution semantics which resembles biology – we apply a principle of maximal parallelism and introduce the novel concept of bounded execution with overshooting. A number of modelling solutions are demonstrated using the example of the well-studied *C. elegans* vulval development process. To date our model is still under development, but first results, based on Monte Carlo simulations, are promising.

1 Introduction

Systems biology [1] is a relatively new field of study which focuses on interactions within and between biological systems. The knowledge about those systems typically is presented in the form of descriptive text, illustrated with diagrams that are often beset with arrows, colourful components and comments. It is not only difficult to locate a particular piece of information, but also to understand it, as there are often unknowns and ambiguities in the description. This problem is inevitable when representing dynamical and concurrent processes in a living cell as flat diagrams.

Furthermore, the amount of biological knowledge is increasing, and has reached the point where the help of machines is becoming indispensable. What is more, *in vitro* (laboratory) experiments tend to be expensive and slow and are often infeasible, whereas *in silico* (computer) experiments could be cheaper, faster and better reproducible. Realistic executable models of biological systems can be used for predictions, preparation and elimination of unnecessary, dangerous or unethical laboratory experiments. This approach would also be applicable, for example, in drug design and testing [2].

J. Fisher (Ed.): FMSB 2008, LNBI 5054, pp. 48–62, 2008.

Therefore, one of the major questions systems biology is currently trying to answer, is how to represent biological knowledge concisely, unambiguously, without omissions, with well-localised gaps and in a machine-processable way.

There are two typical approaches to cell modelling: the first uses systems of differential equations and the second uses stochastic simulations, see for example [3, 4, 5]. They are both quantitative and highly dependent on kinetic constants or reaction rates which are approximate (Sackmann et al. [6] claim that often only 30-50% of data is known). Differential equations work only in cases when many molecules of each protein species are present, and stochastic simulation works only for ,,well-stirred" chemical soups. However, biological cells are far from being ,,well-stirred" and examples abound where small amounts, or even single molecules, are crucial to biological processes. In contract, cell interaction data are available in large amounts [6, 7]. Clearly, there is a need for qualitative rather than quantitative modelling.

Petri nets are a well-established technique for modelling concurrent systems. They are simple and powerful in expressing biological knowledge, e.g. binding, signalling, concurrency, nondeterminism, timing. They are extensible and have intuitive visualisation.

As a model organism we picked a well-described and relatively uncomplicated worm, C. elegans, and the first phase of its well-studied vulval development process [8]. Additionally, this process has discrete output, which makes it easier to verify correctness of a model.

We started with a well-known basic approach to modelling biological systems using Petri nets, described, for example, in [9]. Throughout the development of our model, we set the following three objectives. (1) Resemble biology. (2) Keep the model homogeneous and simple. (3) Comply to the standard Petri net theory. (4) Keep the model qualitative while trying to reproduce the results of laboratory experiments. These goals quite often conflict and it was necessary to find a consensus between them. As a result, we place our work in-between qualitative and quantitative approaches.

In this paper we present our experiences in modelling the C. elegans vulval development process using Petri nets. Instead of extending the formalism with ,,colours" or rates, as is typically done, we focused on preserving the simplicity of the formalism and changing the execution semantics. Rather than the traditional interleaving execution, we use the maximal parallelism principle and introduce a novel concept of bounded execution with overshooting.

Our network is considerably larger than those typically published in papers on modelling of biological cells using Petri nets – it has about 300 places and 300 transitions and approximately 950 edges. Our model is still under development, but first results, based on Monte Carlo simulations, are promising. So far we are able to correctly simulate 46 out of 48 experiments from [10].

The paper is organised as follows. Section 2 sketches the biological process that we are modelling. Section 3 introduces general Petri net theory. Section 4 presents biological modelling methods and execution semantics. Section 5 describes good practices and techniques in modelling using Petri nets. Section

6 identifies problems we encountered while working on the model. Section 7 overviews related work. Section 8 discusses current results. Section 9 gives conclusions and ideas about continuing this work in future.

2   *C. elegans* Vulval Development

C. elegans is a well-studied nematode (type of worm), living in soil and about 1 mm long. It consists of about 1000 cells which are all numbered, and their destinies at all stages of the development of the worm are well-described. Six of those cells are called Vulval Precursor Cells and numbered P3.p, P4.p, P5.p, P6.p, P7.p and P8.p. Those cells form a line. They participate in the process of vulval development, depicted in Fig. 1.

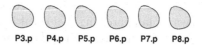

P3.p   P4.p   P5.p   P6.p   P7.p   P8.p

Fig. 1. Cells taking part in vulval development process

Instructions providing all the information necessary for a living cell to grow and function are present in the form of DNA molecules. The DNA contains a number of genes which are templates for the production of proteins, the cell building blocks. We say that a gene is *wild-type (on)* when the protein is normally produced. It is *knocked-out (off)* when it is not produced, and it is *over-expressed* when it is produced in excess. Proteins present in a cell may have two forms, *active* and *inactive*. Typically, the protein must first be activated to participate in reactions. Proteins may also influence other proteins or themselves. For instance, processes of increasing or decreasing the production of a protein are called, respectively, *up-* or *downregulation*. Certain proteins and their interactions are grouped as *pathways*, „packages" with a coarse-grained function in a living cell.

The vulval development process consists of two interacting pathways. Their significant genes are lin-12 and mpk-1. In this process, all cells have to arrive at a decision which fate to choose: 1, 2 or 3. The cells that choose first fate will divide and create the actual vulva. The cells that choose the second fate become supporters of the vulva. The cells that choose the third fate will fuse with the hypodermis, the „skin" of the worm. Before the process starts, all cells are the same. Provided that all genes are produced normally, the process works as follows (see Fig. 2). A special cell, called the Anchor Cell (AC), comes near cell P6.p and induces a signal leading to first fate in P6.p. Then P6.p sends signals to its sides (lateral signals) resulting in the second fate taken by P5.p and P7.p. Remaining cells do not get any signal and eventually choose the third fate. In case the genes were not produced normally, many other possible patterns of fates can occur. Note that this is an extremely simplistic description. This and much more information about *C. elegans* can be found in the Wormbook [8].

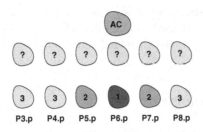

Fig. 2. First stage of *C. elegans* vulval development process. The resulting wild-type fate pattern is (3 3 2 1 2 3).

In this paper we use the following convention: genes are written in lowercase and corresponding proteins in uppercase. The suffix PRO, for example in LIN-12 PRO, denotes the production reaction and suffix DOWN denotes down-regulation. A protein name containing cell number, for example LIN-12/P6.p, denotes protein concentration within this particular cell.

3   Petri Nets

Petri nets [11, 12] are a formalism geared towards modelling and analysis of concurrent systems. A *Place-Transition (PT) Petri net* is a quadruple $(\mathcal{P}, \mathcal{T}, \mathcal{F}, m)$, where $\mathcal{P}$ is a set of places and $\mathcal{T}$ a set of transitions. $\mathcal{F}$ describes weights of arcs which can connect places with transitions or transitions with places. Each place holds zero or more tokens, which represent flow of control through this place. The number of tokens in all places is called a marking of the network, $m: \mathcal{P} \to \mathbb{N}$, and represents its state.

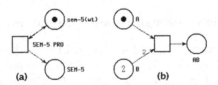

Fig. 3. (a) Production of a protein SEM-5 in the presence of coding gene sem-5(wt). (b) Synthesis of a complex AB from one A and two B.

An example of a Petri net is given in Fig. 3(b). Places are depicted as circles, transitions as rectangles, arcs as arrows and tokens as dots. If no $\mathcal{F}$-value is given, it is 1.

Incoming arcs at a transition represent its requirements and outgoing arcs represent token production. A transition is enabled when all of the requirements are met. An enabled transition can fire: consume all required tokens and produce new tokens. The interleaving execution semantics of a Petri net is defined as follows: in each step select randomly one enabled transition, fire it, repeat.[1] If there are no more enabled transitions left, the network deadlocks. This semantics describes totally asynchronous behaviour, i.e. all possible interleavings of transitions.

4   Modelling Biology

Below we present a method to represent biological knowledge as Petri nets. We illustrate it with three examples of typical biological modelling problems. We also introduce the design of our model and explain the adaptation of the Petri net standard that we have developed.

4.1   Basic Translation

The translation of places and transitions into biological entities is straight-forward. Places represent genes, protein species and complexes. However, we have encountered many cases when we had to represent a single entity with various characteristics as multiple places. For example, to differentiate between active and inactive LIN-12 proteins, we used two places LIN-12 and LIN-12 ACT. Transitions represent reactions or transfer of a signal. Arcs represent reaction substrates and products. Firing of a transition is execution of a reaction: consuming substrates and creating products.

4.2   Gene Expression

Figure 3(a) depicts a typical example of production of proteins from a gene. In this case the „reaction" SEM-5 PRO produces proteins SEM-5 when the wild-type gene sem-5(wt) is present. When the gene is not present, the „reaction" does not take place.

4.3   Downregulation Through Production Suppression

Figure 4 depicts downregulation of a protein LIN-12 through suppressing the expression of a gene lin-12(wt). Normally, if MPK-1 is not present, the reaction LIN-12 PRO is enabled and produces protein LIN-12, similarly to the previous example. However, when MPK-1 is present, the reaction LIN-12 DOWN is enabled and has 0.5 chance of firing compared to LIN-12 PRO, so the production of LIN-12 will halve.

4.4   Downregulation Through Product Removal

Figure 5 illustrates two processes. The first process is another model for decreasing the level of LIN-12 by MPK-1 – the degradation through endocytosis. Here the produced LIN-12 is removed, while in the previous example its

---

[1] In this paper, by random selection we mean that the choice is non-deterministic and all possible choices have equal probability to be selected.

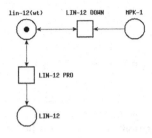

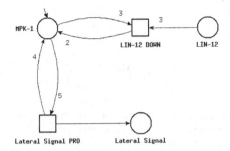

Fig. 4. MPK-1 downregulates LIN-12 by suppression of lin-12(wt) gene

Fig. 5. MPK-1 degradation of LIN-12 through endocytosis and upregulation of a lateral signal by MPK-1

production was reduced. Normally, if MPK-1 is not present, the level of LIN-12 does not change. However, when levels of LIN-12 and MPK-1 are high enough (≥ 3), the reaction LIN-12 DOWN can execute and decrease the concentration of LIN-12. The second process is sending a lateral signal. When a high (≥ 5) level of concentration of MPK-1 is present, a lateral signal can be initiated.

4.5 Concentration Levels

The number of tokens in our model does *not* represent directly the number of molecules of proteins. In the Petri net model, we interpret it in two ways. In case of a gene: 0–not present and 1–present. And in case of a protein: 0–not present and 1-2, 3-4 and 5-6 – low, medium and high concentrations. The rationale behind this approach is to abstract away from unknown absolute molecule concentration levels, in order to keep the model qualitative.

4.6 Maximal Parallelism

The interleaving semantics of Petri nets describes asynchronous behaviour, cf. Sect. 3. However, this is not realistic for biological cells, where all reactions can happen in parallel. Thus, in our model we use the maximal parallelism execution semantics, which can be summarised informally as *execute greedily as many transitions as possible in one step*. In this semantics, a step S is a multi-set of transitions, i.e. a transition can occur multiple times in S. A maximally parallel step is a step that leaves no enabled transitions in the net. A maximally parallel step is enabled if executed transitions are enabled with the required multiplicity. In case more than one maximally parallel step is possible, the one to execute in a simulation is selected randomly. Figure 6 illustrates an example network and possible maximally parallel steps. The maximal parallelism principle for Petri nets is described for example in [13].

Efficient Implementation of Maximal Parallelism. In the implementation of the maximally parallel execution of a Petri net, the key notion is a *conflict*.

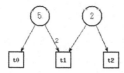

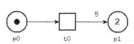

Fig. 6. A network with exactly three possible maximally parallel steps: $\{t0 \times 5, t2 \times 2\}$, $\{t0 \times 3, t1, t2\}$, $\{t0 \times 1, t1 \times 2\}$.

Fig. 7. A bounded network with a bound 6. Transition t0 cannot fire because it would create 7 tokens in p1 and maximally 6 tokens are allowed. Saturation in place p1 is not possible.

We say that two transitions t and t' are *directly in conflict* if they have a common parent. They are *in conflict* if there exists a sequence of transitions $t = t_1, t_2, .., t_k = t'$ such that each pair (t_i, t_{i+1}) is directly in conflict. For example, in Fig. 6 all three transitions are in conflict.

The computational cost of generating the next maximally parallel step is at least $\Omega(\exp(|\mathcal{T}|))$, as it involves verifying all subsets of $\mathcal{T}$ in the worst case, i.e. when all transitions are in conflict. Fortunately, in our experience so far, in nature the networks tend to be ,,sparse" and, as a result, this computational cost has not been a bottleneck.

Algorithm 1. nextMaxStep($\mathcal{N}$: PetriNet)

1: $\mathcal{S} = \emptyset$
2: $\mathcal{D} = \text{generateConflicts}(\mathcal{T})$
3: **for all** $d \in \mathcal{D}$ **do**
4:     $A_d = \text{backtrackAllMaxSteps}(d)$
5:     $s_d = \text{selectRandom}(A_d)$
6:     $\mathcal{S} = \mathcal{S} \cup s_d$
7: **end for**
8: **return** $\mathcal{S}$

Algorithm 1 shows how the next maximally parallel step is created in our implementation. The algorithm proceeds in three stages. In the first stage, the set of all transitions, $\mathcal{T}$, is divided into a set of disjoint subsets of transitions in conflict, $\mathcal{D}$. Existence of enabled transitions in conflict implies that there exists more than one possible step.

In the second stage, for each subset of transitions in conflict, $d \in \mathcal{D}$, all possible maximally parallel steps are generated using a simple backtracking technique. The resulting step, s_d, is then chosen randomly. Note that s_d is a multi-set of transitions that belong to d. In the third stage, the global step is combined out of ,,local" steps: $\mathcal{S} = \bigcup_{d \in \mathcal{D}} s_d$. This can be done because transitions that belong to different subsets in $\mathcal{D}$ are not in conflict. In our implementation, the first stage of partitioning the network into disjoint transitions sets in conflict is executed once, stored and used for creation of a sequence of maximally parallel steps.

4.7 Bounded Execution

In our first attempt, the execution of the network was not bounded, as it was designed to produce the output into infinity. However, it turned out that numerous nodes reached a high number of tokens, e.g. 800, which was in practice impossible to downregulate or degrade. Furthermore, such a high production of certain proteins is not realistic, as in nature the cell would saturate with the product, and the reaction would slow down or stop. Therefore we introduced a bounded execution with a chosen bound $N = 6$, which means that a place cannot contain more than N tokens.

4.8 Bounded Execution with Overshooting

Bounded execution suffers from two problems. First, it is not possible to execute partial reactions, and therefore sometimes saturation cannot be reached – see the example of a network in Fig. 7.

Second, the partition of transitions used for fast execution (see Sect. 4.6) does not work anymore, as there are more conflicts and the entire network becomes inter-dependent. To overcome both problems, we implemented bounded execution with overshooting. Each transition can overshoot maximally once. In other words, a reaction can produce if all products have at least one free token slot in the output place. Consequently, given a place with n incoming arcs with weights $w_1, \ldots, w_n$, the maximal number of tokens in p is $(N-1)+\sum_{i=1}^{n} w_i$.

5   Modelling Solutions

While creating the model of *C. elegans* vulval development process, we developed a number of good practices and techniques for modelling biological knowledge as Petri nets, without extending the formalism. Below we present several of these techniques.

5.1 Reactions Ordering

The fragment of the network depicted in Fig. 5 represented both downregulation of LIN-12 by MPK-1 and sending a lateral signal (see Sect. 4.4). Let us assume that only one token at a time flows into MPK-1. Note that the downregulation of LIN-12 will take place *before* sending a lateral signal. This is because whenever new tokens appear in MPK-1, they first reach the level 3. Only when LIN-12 level drops to < 3 and blocks LIN-12 DOWN, can MPK-1 reach 5 and initiate a lateral signal. The vulval development of *C. elegans* is initiated by the Anchor Cell, as we described in Sect. 2. The P6.p cell receives a strong signal from AC, while its adjacent cells P5.p and P7.p receive weaker signals and an additional lateral signals from P6.p, see Fig. 8.

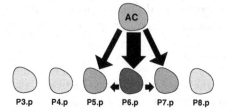

Fig. 8. Anchor Cell initiates the vulval development process by sending stronger and weaker signals

5.2 Modelling Signal Strength

At first we modelled a stronger signal naively – by sending more tokens: 3 instead of 1, as illustrated in Fig. 9. However, this did not work, as our network, unbounded at that time, would just accumulate high concentrations of certain proteins, and, as a result, there was no difference between weak and strong signals. We increased the probability of production of LIN-3/P6.p with a technique depicted in Fig. 10: using multiple transitions for LIN-3/P6.p. Now the probability of executing LIN-3/P6.p is three times higher than the probability of executing LIN-3/P5.p or LIN-3/P7.p. After applying this change, the pathway behaved like a fast „pipeline". Note that this way, spatial vicinity has been translated in the model as the probability of execution.

5.3 Modelling Signal Strength Directly

The method presented in Fig. 10 worked so well that it gave us an idea of directly implementing a transition attribute called **strength**, see Fig. 11. This way we gained clarity in the graphical representation and, additionally, speedup in the maximally parallel step computation, resulting from the reduction of the number of subsets of transitions.

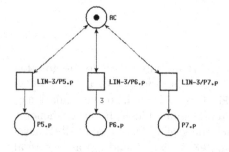

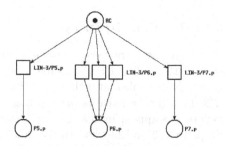

Fig. 9. A simple model for the Anchor Cell signal. Signal strength is modelled as higher production.

Fig. 10. A method to increase the probability of transition execution

The execution semantics needed to be adjusted: instead of randomly selecting a maximally parallel step out of all generated possible steps, we compute a probability that a maximally parallel step should be selected. This probability is the sum of strengths of transitions in the step, normalised by the sum of strengths of all transitions. For example, in Fig. 11, the probability that the LIN-3/P6.p transition should be fired is $\frac{3}{1+3+1} = 0.6$. The realisation of this idea was straightforward to implement.

5.4   Modelling Signal Slowdown

After we switched to modelling with saturation, we observed the following phenomenon: the strong AC signalling pathway in the cell P6.p (see Sect. 5.2) would stall, as most of the places were oversaturated. To overcome this problem we used a method depicted in Fig. 12. In the first step, the transition LIN-3 PRO executes. It results in production of LIN-3 down the pathway, but also puts one token in the cycle. In the next two steps, the signal is not generated, but the token continues cycling. After three steps, the signal is back in AC and the process will repeat. Note that the cycle works as a delay or a buffer: the LIN-3 signal will be created every three steps.

By using this method, we reduced the number of tokens sent within the signal, so that the pathway is able to handle the ,,bandwidth''.

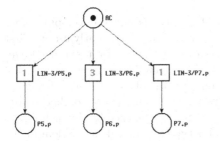

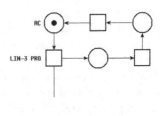

Fig. 11. Transition strength. LIN-3/P6.p is three times stronger than LIN-3/P5.p.

Fig. 12. A method to model signal slowdown. In the first step, the AC signal fires. Afterwards it signals every three steps.

5.5   Good Practice: One Token Moves

In our first attempts to prototype the system, we frequently manually adjusted weights on the arcs. However, this introduced uncertainties and guessed numbers into the model. Thus, we decided to abstract away from them and move just one token wherever possible. (Note, however, that we kept the reaction requirements. If a requirement is k tokens, then we move back $k - 1$ tokens.)

5.6   Unbounded Places

We found that in one important case the places should not be saturated, namely in the environment of the cell. Note, however, that the presence of unbounded places makes the analysis of the net, for example steady state determination, more difficult.

5.7   Model Correctness

Deciding whether a model is correct is difficult and varies from case to case. The *C. elegans* vulval development process has the advantage of discrete output, and for verification we have used 48 experiments published by Fisher et al. [10].

Since the semantics of execution is probabilistic, it is not enough to run one simulation of the network. To verify its behaviour, we ran a Monte Carlo simulation – average multiple runs, typically 3000. We used the DAS-3 cluster [14] for that. For each simulation, we set up genes appropriately and execute the network for a fixed number of steps.

Our observations show that protein concentrations reach „steady" levels after a fixed number of steps. We we use these levels to determine cell fates and check whether the output is correct. Our experience so far is that single runs of the simulation are good representatives of average behaviour, that is to say standard deviation of network behaviour is small.

5.8   Debugging Simulations

In our attempts to create a correct model of *C. elegans* vulval development, we found two approaches to debugging Petri net simulation particularly useful. First are the plots of protein concentration levels during a simulation, averaged over various windows. Second, we use a backtracking technique. During simulation we generate statistics about the average numbers of tokens in places, firings of transitions and executed steps. The statistics aid us to manually go „backward" from output places, and trace which transitions were blocked, and which places were oversaturated or depleted. This method typically reveals a problem in the model, for example too strong downregulation or an incorrect concentration level, and gives good hints as to why the execution did not work as it was supposed to.

6   Problems Identified

While extensively working with large Petri nets we noted several disadvantages of using this formalism for modelling biological cells.

6.1   Drawing Tool

A major practical problem when creating the biological model is the lack of a graphical tool supporting handling of large networks (by large we mean more than 100 nodes). The most desirable features of such a tool are:

- Collective operations such as labelling, moving, hiding.
- Clever zooming, for instance zooming into a subset of the network.
- Adding new types of labels to nodes. For example, we needed to add label *strength* to transitions, and we ended up using non-portable ways of doing that.
- Support for the Petri Net Markup Language [15]. Using this standard turned out very convenient, as we were able to switch drawing tools multiple times.
- Advanced visualisation of execution, allowing, for example, to view only the selected nodes and switching of the viewing modes.

TINA [16] is the tool that we had best experiences with.

6.2   Modularity, Compositionality

Another disadvantage of using Petri nets for biological modelling is that they offer little support for modularity and compositionality. Our model consists of six identical cells, and each cell consists of two pathways. Applying a modular approach would be very useful in this case. Also, with the view of possibly including several development stages of *C. elegans* in the future, applying a modular approach would be indispensable.

6.3   Synchrony vs. Asynchrony

Biology is neither totally synchronous nor totally asynchronous. For example, a chemical reaction, i.e. creation of products and consumption of substrates, can be thought of as immediate, synchronous. By contrast, sending a signal to another cell is a typical asynchronous operation, like sending a letter. And last but not least, time and quantities play an important role, meaning that communications in different parts of a biological system tend to occur at the same rate. A formalism to model biological systems should be able to express such dependencies.

Petri net interleaving semantics models totally asynchronous behaviour (cf. Sect. 3). We switched to a maximally parallel semantics (cf. Sect. 4.6), which progresses in lockstep, i.e. it models synchronous behaviour. In our experience, this way we achieve a much closer resemblance to biological cells.

However, we lost the possibility to model asynchronous communication. An important open question is: how to introduce asynchrony into a lockstep? Bounded asynchrony is discussed in [17], but has a disadvantage of incorporating a global timer, and [18] focuses on coordination orchestration, which seems heavy-weight for cell modelling.

7   Related Work

Petri nets were introduced by C. A. Petri in his dissertation in 1962 [11]. An article by Murata [12] contains a good introduction to general Petri net theory. Reddy was the first to use Petri nets for modelling of biochemical reactions. He presented a basic translation of a biological narrative into a Petri net and an analysis with invariants [9]. Since that time, numerous papers applying Petri nets to biological modelling were published, for example [6, 19, 20, 21]. All of these papers use interleaving execution semantics and do not model saturation. Resulting nets are also much smaller than ours. A different approach take Genrich et al. [2], who create their network by querying biological databases, such as KEGG [22]. The amount of data is huge and the resulting network is at first gigantic, but then simplified to only 8 medium-size pathways.

Many extensions to Petri nets have been developed and used in modelling of biological systems: Coloured Petri Nets (tokens have colours) [2], Stochastic Petri Nets (transitions have rates) [21,23], Timed Petri Nets (transitions have delays) [20], Hybrid Petri Nets (places can be continuous) [24], Functional Petri Nets (transitions have functions) [25], and Hybrid Functional Petri Nets (everything mixed) [26]. There is also a recurring approach of representing a biological pathway in logic and translating it automatically into a Petri net [6, 27, 28]. The reader may find recent survey papers concerning modelling of biological systems in [7, 26, 29].

The most common way of validation of Petri net models of biological systems are simulations and comparing the evolution of concentration levels or reached steady state to data in the literature, e.g. [20]. More advanced structural analysis involves determining invariants or checking for deadlocks [9, 19] and model checking, e.g. [28].

Besides Petri nets, there are numerous formalisms to model and analyse biological pathways, such as process calculi, Boolean networks, statecharts, REO. The reader might consult [18, 26, 30]. It is still not clear though which of those techniques are best suited for biological modelling.

8   Current Results

At the time of writing this article, our model is still under development. Our network is considerably larger than those typically published in papers on modelling of biological cells using Petri nets – it has about 300 places and 300 transitions and approximately 950 edges. We localised a lot of the necessary biological information in the literature, for example [8, 31, 32].

First results look promising. We developed an implementation of the execution method, and are able to correctly repeat 46 out of 48 experiments from [10], using Monte Carlo simulations.

9   Conclusions

In this paper we presented the technique of modelling biological cells as distributed systems represented by Petri nets. We base our findings on the *C. elegans* vulval development model that we have built. We propose to model with the basic Petri net formalism, which is easy to understand and visualise, and to adapt its execution semantics to resemble biology. Our execution semantics is based on the maximal parallelism principle (execute in parallel as much as possible) and bounded execution with overshooting (saturation).

Petri net models that we use are both qualitative and quantitative. Qualitative – because network structure represents static knowledge about interactions within a cell. Quantitative – because places and transitions can contain or require multiple tokens and we exploit this to express different signal strengths and concentration levels. For instance, we can express that one protein is produced five times faster than another protein, or that a concentration level for downregulation has been reached.

There is currently a need for executable biological models, but no consensus on what is the best biological modelling language [30]. In our experience so far, quantitative Petri nets with maximal parallelism and bounded execution could be a good choice. They very naturally feature concurrency, nondeterminism, synchronisation and visualisation, they allow to model systems that resemble the actual behaviour of biological cells, and a Monte Carlo simulator can be implemented efficiently.

In the near future we intend to work on issues such as modularity, synchrony versus asynchrony, system robustness, steady states, and model checking.

References

[1] Alon, U.: An Introduction to Systems Biology: Design Principles of Biological Circuits. Chapman & Hall/CRC (2006)
[2] Genrich, H., Küffner, R., Voss, K.: Executable Petri net models for the analysis of metabolic pathways. International Journal on Software Tools for Technology Transfer (STTT) 3(4), 394–404 (2001)
[3] E-Cell, http://www.e-cell.org
[4] Gillespie, D.T.: Exact stochastic simulation of coupled chemical reactions. The Journal of Physical Chemistry 81(25), 2340–2361 (1977)
[5] Srivastava, R., You, L., Summers, J., Yin, J.: Stochastic vs. deterministic modeling of intracellular viral kinetics. Journal of Theoretical Biology 218(3), 309–321 (2002)
[6] Sackmann, A., Heiner, M., Koch, I.: Application of Petri net based analysis techniques to signal transduction pathways. BMC Bioinformatics 7, 482 (2006)
[7] Chaouiya, C.: Petri net modelling of biological networks. Briefings in Bioinformatics 8(4), 210–219 (2007)
[8] Wormbook, http://www.wormbook.org
[9] Reddy, V., Mavrovouniotis, M., Liebman, M.: Qualitative analysis of biochemical reaction system. Computers in Biology and Medicine 26(1), 9–24 (1996)
[10] Fisher, J., Piterman, N., Hajnal, A., Henzinger, T.A.: Predictive modeling of signaling crosstalk during C. elegans vulval development. PLoS Computational Biology 3(5), 92 (2007)
[11] Petri, C.A.: Kommunikation mit Automaten. Schriften des IIM Nr. 2, Institut für Instrumentelle Mathematik, Bonn (1962)
[12] Murata, T.: Petri nets: Properties, analysis and applications. Proceedings of the IEEE 77(4), 541–580 (1989)
[13] Kleijn, J.H., Koutny, M., Rozenberg, G.: Towards a Petri net semantics for membrane systems. In: Freund, R., Păun, G., Rozenberg, G., Salomaa, A. (eds.) WMC 2005. LNCS, vol. 3850, pp. 292–309. Springer, Heidelberg (2006)

[14] Distributed ASCI Supercomputer DAS-3, http://www.cs.vu.nl/das3

[15] Petri Net Markup Language, http://www.informatik.hu-berlin.de/top/pnml

[16] TINA, http://www.laas.fr/tina

[17] Fisher, J., Henzinger, T.A., Mateescu, M., Piterman, N.: Bounded asynchrony: A biologically inspired notion of concurrency. Technical report, MTC (2007)

[18] Clarke, D., Costa, D., Arbab, F.: Modelling coordination in biological systems. In: Margaria, T., Steffen, B. (eds.) ISoLA 2004. LNCS, vol. 4313, pp. 9–25. Springer, Heidelberg (2006)

[19] Koch, I., Junker, B.H., Heiner, M.: Application of Petri net theory for modelling and validation of the sucrose breakdown pathway in the potato tuber. Bioinformatics 21(7), 1219–1226 (2005)

[20] Li, C., Ge, Q.W., Nakata, M., Matsuno, H., Miyando, S.: Modelling and simulation of signal transductions in an apoptosis pathway by using timed Petri nets. Journal of Biosciences 32, 113–127 (2007)

[21] Goss, P., Peccoud, J.: Quantitative modeling of stochastic systems in molecular biology by using stochastic Petri nets. Proceedings of the National Academy of Sciences of the United States of America (PNAS) 95(12), 6750–6755 (1998)

[22] KEGG, http://www.genome.jp/kegg

[23] Srivastava, R., Peterson, M.S., Bentley, W.E.: Stochastic kinetic analysis of the Escherichia coli stress circuit using σ-32-targeted antisense. Biotechnology and Bioengineering 75(1), 120–129 (2001)

[24] Matsuno, H., Doi, A., Nagasaki, M., Miyano, S.: Hybrid Petri net representation of gene regulatory network. In: Proceedings of Pacific Symposium on Biocomputing, vol. 5, pp. 341–352 (2000)

[25] Hofestädt, R., Thelen, S.: Quantitative modeling of biochemical networks. In Silico Biology 6, 39–53 (1998)

[26] Matsuno, H., Li, C., Miyano, S.: Petri net based descriptions for systematic understanding of biological pathways. IEICE Transactions on Fundamentals of Electronics, Communications and Computer Science E89-A(11), 3166–3174 (2006)

[27] Simao, E., Remy, E., Thieffry, D., Chaouiya, C.: Qualitative modelling of regulated metabolic pathways: application to the tryptophan biosynthesis in E. Coli. Bioinformatics 21(2), 190–196 (2005)

[28] Steggles, L., Banks, R., Wipat, A.: Modelling and analysing genetic networks: From Boolean networks to Petri nets. In: Priami, C. (ed.) CMSB 2006. LNCS (LNBI), vol. 4210, pp. 127–141. Springer, Heidelberg (2006)

[29] Peleg, M., Rubin, D., Altman, R.B.: Using Petri net tools to study properties and dynamics of biological systems. Journal of the American Medical Informatics Association (JAMIA) 12(2), 181–199 (2005)

[30] Fisher, J., Henzinger, T.A.: Executable cell biology. Nature Biotechnology 25(11), 1239–1249 (2007)

[31] Yoo, A.S., Bais, C., Greenwald, I.: Crosstalk between the EGFR and LIN-12/Notch pathways in C. elegans vulval development. Science 303(5658), 663–666 (2004)

[32] Shaye, D.D., Greenwald, I.: Endocytosis-mediated downregulation of LIN-12/Notch upon Ras activation in Caenorhabditis elegans. Nature 420(6916), 686–690 (2002)

Combining Intra- and Inter-cellular Dynamics to Investigate Intestinal Homeostasis*

Oksana Tymchyshyn[1] and Marta Kwiatkowska[2]

[1] School of Computer Science, University of Birmingham, Birmingham B15 2TT, UK
[2] Oxford University Computing Laboratory, Wolfson Building, Parks Road, Oxford OX1 3QD, UK

Abstract. This paper reports on the multi-scale modelling of an intestinal crypt cellular structure coupled with Wnt signalling. Using formal modelling techniques based on the stochastic π-calculus, which supports ambients needed for compartments, we develop a collection of cell and molecular level models. The focus of our study is the role of Wnt in the control of cell division and differentiation. Using the BioSPI simulation platform, we analysed the model and reveal a plausible explanation for a mechanism that ensures robustness of cell fate determination.

1 Introduction

Analysis of signal transduction has uncovered a remarkable complexity of signalling networks in terms of their structure and dynamics. Despite the availability of a vast amount of data on the properties of individual molecules, the understanding of the function performed by a particular molecular network as a whole is still lacking. The complexity of signalling architecture can be explained by the need to optimize cellular response to the information it receives about environmental and internal conditions. In particular, a significant degree of this complexity can be attributed to ensuring robustness of the response despite varying environmental conditions.

Numerous experimental data are known that implicate the Wnt pathway in the control of different aspects of homeostasis [1,2,3,4,5]. It has been proposed that Wnt signalling regulates cell proliferation and renewal in the large intestinal epithelium. The large intestinal tract is built of geometrical tubular structures, called crypts. Intestinal homeostasis involves cell generation by division at the crypt base, progressive cell differentiation while they migrate to the top of the crypt, and cell death followed by extrusion when they reach the top. Stem cells, believed to reside at the crypt bottom, have the unique ability to maintain the entire epithelium. As they divide and move up, stem cells must constantly adjust their behaviour by entering partially differentiated population (called transit) prior to terminally differentiating. Simultaneously, the proliferative capability of transit cells is the highest and decreasing as cells move upwards. The question

---

* Supported in part by EPSRC grant GR/S72023/01.

J. Fisher (Ed.): FMSB 2008, LNBI 5054, pp. 63–76, 2008.

that arises is which factors control the ability of intestinal cells to keep a fine-tuned balance between cell division and differentiation.

In this paper, we test the feasibility of different biological hypotheses about the influence of Wnt signalling on the cell fate and the emergence of the robust regulation of cell numbers in the tissue. The function of Wnt cannot be measured experimentally; rather, only the average behaviour of the collection of cells in response to Wnt factors can be observed. We therefore employ computational modelling to examine if the specific properties of the Wnt pathway architecture can provide the conditions for the emergence of the robust regulation mechanism that ensures homeostasis in the intestine.

We base our approach on a conceptual extension of the stochastic π-calculus for spanning multiple scales. We describe in detail how to build a multi-scale model that couples signal transduction network to cellular decisions to proliferate and differentiate. We then analyze the model to reveal how a population of cells interacts and develops into a tissue under the influence of the environment.

2   Related Work

Several modelling approaches for studying the self-renewal process in the intestine exist [7,8,9]. A recent model [7] is representative of the class of deterministic spatially-uniform models. The authors describe the evolution of cell numbers in stem, transit, and differentiated compartments, assuming the constant compartment-dependent rates of renewal, differentiation and death. The model is shown to be very sensitive to changes in these macroscopic rate constants. The authors subsequently investigate the impact of the hypothetical negative feedback mechanism that, based on regulation of the rate at which cells differentiate, allows the crypt to maintain the equilibrium in cell numbers.

In a similar compartment-based but stochastic approach [8], crypt growth is described by a Markov process that models a stem cell population in which each stem cell produces zero, one, or two stem cells, according to a fixed probability distribution that does not vary from cell to cell. In the same manner as [7], the probability of self-renewal vs differentiation is assumed to be pre-programmed and independent of the conditions, except in case of stem cells knowing their number. Both models, however, do not indicate how the knowledge of stem cell numbers is acquired and propagated between physically separated cells. No experimental evidence exists that supports this assumption.

The incorporation of spatial cell fate control mechanism is achieved in [9], where deterministic model for crypt proliferation regulated by diffusible growth factor is presented. Epithelium is modelled as a one-dimensional array of cells. Each cell enters a cell cycle only if the growth factor concentration in the respective cell exceeds a certain threshold. Growth factor is spread by diffusion starting from the bottom of the crypt, but the concentration of growth factor in the tissue is constant. The model mechanism ensures the dynamic regulation of cell proliferation without the need to impose a static type-dependent program executed by every cell. However, under more realistic conditions of stochastic time-varying growth factor field, the accuracy of this mechanism would collapse, resulting in high variability in the numbers of proliferative cells and crypt size.

3   Using π-calculus for Modelling Intracellular Dynamics

The process-algebra approaches, originally developed in computer science for describing and executing networks of concurrent components, have since been successfully applied for analyzing molecular and genetic systems [10,11,12]. Once a biological system has been modelled using basic components of the process algebra, the model can be stochastically simulated to derive the properties under study over time. Stochastic π-calculus [6] is one type of process algebras where interactions are assigned rates controlled by exponential distributions. In this paper, we use the BioSPI [11] as the platform which performs simulations of the π-calculus code using an adaptation of the Gillespie algorithm [13].

In this section, we describe how stochastic π-calculus can be applied to modelling and analyzing molecular interactions and transient changes occurring within cell. At the intracellular level, molecules are modelled as independent agents, governed by discrete reaction rules. Concurrent molecular agents are interconnected to describe dynamic changes of intracellular state in response to internal and external changes.

3.1   Molecules as Mobile Processes

A model in the stochastic π-calculus is a composition of concurrent components (called processes), each of which operates as a state machine describing the possible behaviours of the component. Processes communicate by sending data on channels, which they can dynamically create and destroy. Probabilistic choice, parallel composition, and scope restriction are among the built-in primitives of the π-calculus.

A process X defined as a choice

$$X ::= \pi_1, X_1 + \dots + \pi_n, X_n$$

may evolve as either of X_i, depending on which of the operations π_i (explained below) is the first one to complete in the current context, thus representing a race condition between a set of processes. A process X given by

$$X ::= X_1 | \dots | X_n$$

denotes a composition of processes X_1, ..., X_n running in parallel. The creation of the private channel r within the scope of the given process X is achieved by the operator

$$(new\ r(c))\ X$$

where channel r with rate c is bound to process X. Only processes that share a private channel may interact using that channel. Timing is incorporated into

π-calculus models by associating each channel r with the rate governed by the exponential distribution with the mean $1/c$.

Another basic operation of the π-calculus is synchronous communication of a pair of processes over a channel. Output $r!$ and input $r?$ prefixes, where r is a channel name, are elementary constituents of communication capabilities. The result of communication between process $X ::= r!\{y\}, X'$, containing an output capability $r!\{y\}$, and process $Y ::= r?\{z\}, Y'$, containing an input capability $r?\{z\}$, follows from the central reduction rule:

$$r!\{y\}, X'|r?\{z\}, Y' \longrightarrow X'|Y'[y/z],$$

where y is substituted for z in Y'. Communication between processes may carry information that further changes their interaction capabilities.

It is possible to represent π-calculus processes using a graphical notation. In graphical π-calculus, a model is a graph whose nodes correspond to processes and edges correspond to state transitions. Fig. 1 illustrates how basic operations are represented graphically.

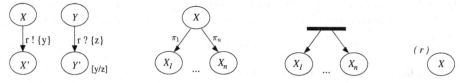

Fig. 1. Communication, choice, parallel composition, and scope restriction primitives of the stochastic π-calculus

A translation scheme that maps molecular reaction networks into π-calculus programs was the subject of previous publications [10,11]. Molecular entities can be coded in the π-calculus as processes that participate in reactions by communicating over channels. State transitions resulting from process communication correspond to covalent modification, association/dissociation, or degradation of signalling molecules. Molecules with several independent functional domains are represented as a parallel composition of π-calculus processes.

3.2   The Model of the Wnt Signalling Pathway

Wnt signalling induces a great variety of cell responses, spanning from morphogenesis and adult tissue homeostasis, to cancer formation. The main event of Wnt signalling is the accumulation of β-catenin which sends a signal into the nucleus for further processing. In the absence of the signal, a cytoplasmic degradation complex consisting of proteins Axin and APC, and kinase GSK3, rapidly degrades β-catenin. To activate the pathway, extracellular Wnt binds to the membrane receptor complex which generates the signal by inhibiting the degradation complex. This reduces the degradation rate of β-catenin, leading to its accumulation in the nucleus [14].

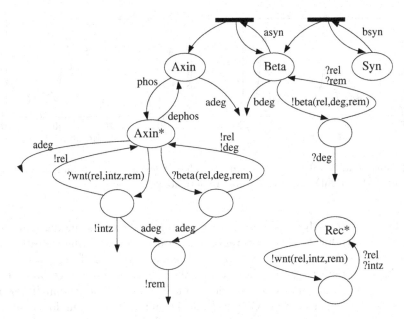

Fig. 2. π-calculus model of the Wnt pathway

The detailed model of the Wnt signalling pathway has been described elsewhere [15,16]. We further experiment with the model by testing additional mechanisms with respect to their capacity to enhance system robustness and adaptability. Altering the number of intracellular components, for example, is expected to alter cellular response, but if the system is robust, the extent of this alteration will be minimized. The system has, however, to sense and adapt to changes of the environmental conditions. We find that the addition of feedback loops [14] has a significant effect on promoting system robustness to parameter variation, a characteristic crucial for reliable performance of many biochemical networks.

Fig. 2 describes the π-calculus implementation of the revised Wnt pathway model. The process *Beta*, a π-calculus abstraction of β-catenin, interacts on channel *beta* with the process *Axin** used to represent an activated state of the destruction complex. After interaction is completed, the resulting process can use channel *rel* to break the binding and return to its original state *Beta*, or can be degraded (transition marked *deg*). *Axin* can transit between inactive and active states with delay times specified by *phos* and *dephos*. Active *Axin* contains a complementary interaction capability *?beta* that allows it to interact with *Beta*. Using channel *wnt*, *Axin** can bind and subsequently be inhibited by the activated receptor complex *Rec**. A negative feedback loop is created by *Beta* which spawns further instances of the process *Axin* (at the rate specified by *asyn*).

In Fig. 3 we plot model outputs obtained from a single simulation run. Depending on the strength of the incoming stimulus Wnt (represented as a fraction of activated receptors), the system allows for different dynamic regimes. At low levels of Wnt, the model predicts stochastic outbreaks of β-catenin activity

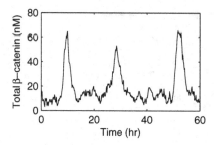 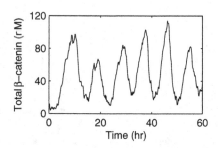

Fig. 3. Steady state value of β-catenin for different levels of the incoming Wnt: 25% (left) and 100% (right) of the receptors are activated

(Fig. 3 (left)) (note that the deterministic counterpart of the model reaches a steady state with low levels of β-catenin, for details see supporting website [25]). As the stimulus increases, the oscillations become regular and are most coherent in the deterministic limit when the number of molecules is very large, thus corresponding to the limit cycle behaviour of the deterministic system. These results agree with experimental evidence that high-intensity staining for β-catenin is observed only in a few cells at the bottom of the crypt, where Wnt signal is the strongest. Transient β-catenin is observed throughout the proliferative compartment in the lower 2/3 of the crypt. Indeed, we validated model predictions of oscillatory pathway dynamics in human cell cultures (a manuscript in preparation).

4 Extending a Framework to Model Cells

We are interested in testing possible hypotheses about Wnt-based control of cell proliferation and differentiation in the intestine. To test the feasibility of the mechanisms proposed by different research groups, we build a model that couples cellular decisions with the state of the Wnt signalling network embedded in every cell. Next, we explain how an extension of the π-calculus can be used for spanning different scales of the biological system.

4.1 Cells as Mobile Ambients

In order to extend the model with the cell-level dynamics, we first acquire a mechanism of embedding molecules into cells. We use ambients [19] to define a bounded place where interactions between agents happen. Enrichment of the stochastic π-calculus with ambients, introduced in [17], provides an ability to specify communication between π-calculus processes based on their location within a common boundary.

An ambient is a location where communication happens: $cell[X]$ stands for the process X running at the location $cell$ (i.e., in ambient $cell$). Locations may reside within locations: in $cell[mol[A] \mid mol[B]]$ two ambients mol are incorporated into

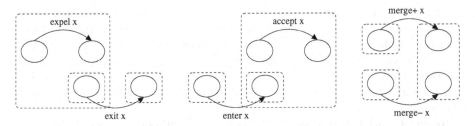

Fig. 4. Ambient capabilities

the ambient *cell*. Computation may contain the reconfiguration of a hierarchy of locations. In the following, we graphically represent an ambient as a dashed rectangle around the processes and sub-ambients it contains, possibly labelled with the ambient name. The derived models are simulated using the BioSPI platform [11] which supports ambients without modifying the semantics of the stochastic simulation [13].

Spatial configuration of the ambient system can be changed using capabilities such as *exit/expel* from the ambient, *accept/enter* into the ambient or *merge* with the ambient (Fig. 4). Communication abstraction is extended to represent compartment restriction on interactions based on their locations. Three types of communication restrictions are *local* (between processes in the same ambient), *s2s* (between processes in sibling ambients) and *p2c/c2p* (between processes in parent and child ambients) (Fig. 5).

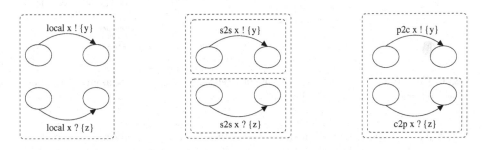

Fig. 5. Communication directions between ambients

To represent cells, allowing molecules to be assigned and re-assigned to specific cells, we abstract cells as ambients. Consequently, molecular communication within one cell is abstracted by the *s2s* communication direction. For communicating the state of the intracellular molecular network to the cell decision-making process, we use the *p2c/c2p* direction.

4.2 Modelling Cell Division and Movement

Here we present the spatial abstraction that describes the diffusion of the extracellular morphogene in one direction in the tissue. The pressure exerted by cell division due to higher amounts of the morphogene directs the cell to move away from the morphogene source. This would accommodate the scenario of Wnt morphogene distribution along the crypt length and cell movement to the top of the crypt. Analogous extension of the π-calculus framework with spatial information is necessary when the desired objective is to simulate diffusion of extracellular growth or inhibitory factors, competition for space between different cells, or cell adhesion.

To model spatial abstraction, we define a neighbourhood relationship between cells. Two cells are neighbours if they share a private channel which is used to send instructions from one neighbour to another. In one-dimensional space, it is sufficient for each cell to keep the reference to its upper neighbour (channel *next* in Fig. 6). Extracellular signal and cell movement are functions of the neighbourhood. Following cell division, the upper neighbour is requested to free its position by moving upwards. Another instance of a cell is created and is inserted in the neighbouring position by updating its references to the neighbourhood, as illustrated in Fig. 6.

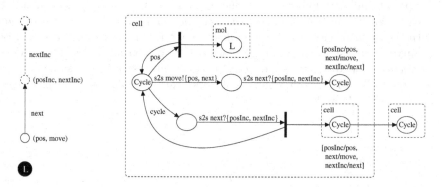

Fig. 6. Cell organization: linear array of cells referencing upper neighbours (left); and lattice representation in π-calculus (right)

Diffusion of the external morphogene is simulated by calculating the concentration of the external factor field at each cell position rather than simulating the movement of factor molecules within the spatial lattice. The channel *pos* with appropriate rate is carried by each cell to indicate its distance to the morphogene source.

5   A Model of Intra- and Inter-cellular Dynamics of the Crypt

Numerous and often inconsistent evidence exists suggesting that Wnt signalling controls the balance between cell proliferation and differentiation in the intestinal crypts and other tissues. Wnt is suggested to influence cell advance or withdrawal from the cell cycle [1,3,4], and cell ability to maintain its stem-cell phenotype or to differentiate [2,5,21]. To test different hypotheses about the Wnt-based regulatory mechanisms involved in the intestinal homeostasis, we build a multi-scale model that couples the state of the intracellular network based on the previously described Wnt pathway to different decisions that the cell might make. The extracellular diffusible Wnt triggers changes in the intracellular state and thus influences cellular behaviour. We examine how these mechanisms influence the robust turnover of cells in the intestinal crypt and its disregulation in cancer.

5.1 Proliferative and Differentiated Cell Fate

In our model, we adopt two threshold mechanisms to decide whether the cell undergoes proliferation, differentiation, or stays quiescent. Increased β-catenin activity influences the initiation of a new cell cycle. The time to complete the cycle is assumed to follow an exponential distribution. Variability of the cycle length is thus incorporated into a delay needed for the cell to make a decision to proliferate.

In addition, β-catenin expression is linked to the ability of a stem cell to preserve its phenotype. We assume that once the cell starts expressing differentiation markers, differentiation is irreversible. While stem cell divides to produce two cells with an equal stem-cell capability, differentiated cell divides to produce two identical differentiated cells. Differentiated cells are also assumed to have a limited life span, as opposed to stem cells which are subjected to only a low-level apoptosis.

The alternative hypothesized scenarios of cell-fate decisions which we compare are:

Hypothesis 1. Transient activation of β-catenin in the cell triggers initiation of a new cell cycle. High levels of β-catenin are required to preserve stem cell properties.

Hypothesis 2. Transient activation of β-catenin is sufficient to push the cell into a new cycle while prolonged β-catenin signalling causes the stem cell to start expressing differentiation markers.

Each threshold mechanism is associated with a π-calculus channel which transmits a signal to the cell once the level of the intracellular β-catenin exceeds a specified threshold. In Fig. 7, channel molcycle is used to instruct the cell to enter a new cycle. Another threshold moldiff blocks (Hypothesis 1) or triggers (Hypothesis 2) cell differentiation (Fig. 7).

5.2 Wnt Gradient in the Tissue

Because Wnt targets are generally expressed in stem and proliferative cell compartment, it is widely accepted that Wnt factors are produced at the bottom of the crypt and are then transported by diffusion [20]. However, it has recently been suggested that Wnt gradient follows a more complex pattern due to surprisingly strong expression of the extracellular Wnt inhibitors at the bottom of the crypt [24]. We approximate this by additionally decreasing the rate at which Wnt is received by cells located at the bottom of the crypt spatial lattice (channel *pos* in Fig. 7).

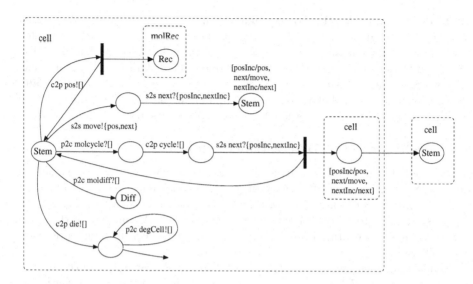

Fig. 7. Stem cell evolution: a cell undergoes proliferation, differentiation, or death. Additionally, the cell is constantly receiving information about the environment, and adjusts its position within the spatial lattice to accommodate newly born cells.

6   Robust Cell Fate Determination by Wnt Signalling

Using the BioSPI platform, we perform extensive simulations of the described scenarios in order to derive the properties of the multi-scale cellular system whose regulatory control is the extracellular diffusible factor Wnt. The derived models are subsequently analyzed with respect to the number of cell divisions as a function of cell position along the crypt vertical axis (i.e., distance to the Wnt source), the total number of cells in the crypt, and influence of stochasticity and random parameter perturbations on the tissue response (details about model parameters used in simulations and full model implementation are available at the website [25]).

The first family of models implement cellular decision mechanism described by Hypothesis 1. Our analysis shows that under these assumptions the fate that the cell assumes is very sensitive to the level of Wnt it is exposed to. The distribution of proliferating cells mimics the distribution of the Wnt factors along the crypt axis. The result is a high variability of the size of proliferative cell compartment and crypt size, which is inconsistent with the experimental observations. Moreover, activating mutations in the Wnt pathway, which increase the level of intracellular β-catenin, lead to significant expansion of the stem cell compartment. Consequently, the number of cells in the crypt becomes unstable and starts growing exponentially. We conclude that Hypothesis 1 is unable to reproduce the tissue response observed experimentally.

Simulations of the model based on Hypothesis 2 reveal that this combination of intracellular and cellular dynamics ensures robust tissue response mediated by

Wnt (Fig. 8 (left)). Rather than being scattered throughout crypt length, proliferative cells are confined to the restricted compartment at the bottom of the crypt. This is consistent with the experimental data ([18,22] and Fig. 8 (right)). The number of proliferative cells as well as the total number of cells in the crypt shows little variability, despite random noise and stochastic perturbations present in the model. This is consistent with the reports of a surprisingly narrow distribution of crypt sizes, the fact that has not yet been reproduced in modelling studies.

We next investigate the effects of the mutations in the Wnt pathway which were identified in concrete cancer models: Familial Adenomatous Polyposis [18], hyperplastic and adenomatous polyps [22,23]. To simulate the effect of mutations, we decrease the rate of the β-catenin inhibition by the active APC/Axin destruction complex (channel *beta* in Fig. 2). Up to 5-fold decrease of the β-catenin inhibition rate results primarily in a shift of the proliferative cells toward the top of the crypt (Fig. 8 (left)). The size of the crypt is increased only slightly. These predictions are in good agreement with the experimental evidence [18,22]. Fig. 8 (right), adapted from [18], shows experimental evidence of changes in the structure of the proliferative compartment resulting from mutations that decrease the activity of β-catenin inhibitor complex.

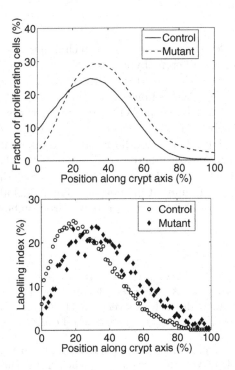

Fig. 8. Cell fate control by the Wnt pathway: model predictions (left) of the proliferative cell distribution in both healthy and mutant tissues agree well with the experimental data (right)

Further inhibition of the destruction complex leads to more advanced forms of intestinal cancer: colorectal adenomas [22,23]. While cell proliferation shifts upwards at the initial stage, the model predicts the break up of the mechanism that confines proliferative cells to the bottom of the crypt. This is consistent with the experimental observations of proliferation in adenomas being almost evenly distributed throughout the crypt length [22,23].

Our model provides an explanation to the observed phenomena. Cell proliferation is triggered by even modest increase of the Wnt levels which is sufficient to upregulate β-catenin to high amplitude. As Wnt increases, stochastic oscillations in β-catenin expression become deterministic and their frequency increases along with the cell proliferation rate. Analysis of the model also shows that under Hypothesis 2, which links stem cell fate to the region of rare stochastic oscillations of β-catenin activity, stem cells are limited to low Wnt region and decrease in numbers under mutant conditions. Thus, the model is not only consistent with the reports of low β-catenin activity in stem cells [2,5] and the reduced proliferation rate of stem cells caused by rare outbreaks of β-catenin, but also suggests the protection mechanism against the stem cell expansion that would immediately lead to the exponential growth of tumours [7].

7   Conclusions

In this paper, we employed formal modelling techniques based on the stochastic π-calculus to examine different hypotheses about the influence of the Wnt pathway on homeostasis of the intestinal epithelium, and its role in tumourigenesis. We proposed that possible function of Wnt is to ensure the robust cell fate determination. The model of the Wnt signalling pathway was subsequently coupled to the cellular behaviour and the environment to test its role in maintaining a fine-tuned balance between cell division and differentiation. The result of the model is consistent with different properties of the distribution of cells in the crypt. The model can explain both the stability of the healthy regulation and the changes seen in mutant phenotypes. The model also suggests which characteristics of tissue architecture can protect it from unbounded growth.

References

1. Wetering, M., Sancho, E., Verweij, C., Lau, W., Oving, I., Hurlstone, A., Batlle, E., Coudreuse, D., Haramis, A., Tjorn-Pon-Fong, M., Moerer, P., Born, M., Soete, G., Pals, S., Eilers, M., Medema, R., Clevers, H.: The beta-catenin/TCF-4 complex imposes a crypt progenitor phenotype on colorectal cancer cells. Cell 111, 241–250 (2002)
2. Dravid, G., Ye, Z., Hammond, H., Chen, G., Pyle, A., Donovan, P., Yu, X., Cheng, L.: Defining the role of Wnt/beta-catenin signaling in the survival, proliferation, and self-renewal of human embryonic stem cells. Stem Cells 23, 1489–1501 (2005)
3. Reya, T., Duncan, A.W., Ailles, L., Domen, J., Scherer, D.C., Willert, K., Hintz, L., Nusse, R., Weissman, I.L.: A role for Wnt signalling in self-renewal of haematopoietic stem cells. Nature 423, 409–414 (2003)

4. Sato, N., Meijer, L., Skaltsounis, L., Greengard, P., Brivanlou, A.H.: Maintenance of pluripotency in human and mouse embryonic stem cells through activation of Wnt signaling by a pharmacological GSK-3-specific inhibitor. Nat. Med. 10, 55–63 (2004)

5. Lowry, W., Blanpain, C., Nowak, J., Guasch, G., Lewis, L., Fuchs, E.: Defining the impact of beta-catenin/Tcf transactivation on epithelial stem cells. Genes. Dev. 19, 1596–1611 (2005)

6. Priami, C.: Stochastic pi-calculus. Comp. J. 38, 578–589 (1995)

7. Johnston, M.D., Edwards, C.M., Bodmer, W.F., Maini, P.K., Chapman, S.J.: Mathematical modeling of cell population dynamics in the colonic crypt and in colorectal cancer. Proc. Natl. Acad. Sci. USA 104, 4008–4013 (2007)

8. Loeffler, M., Bratke, T., Paulus, U., Li, Y.Q., Potten, C.S.: Clonality and life cycles of intestinal crypts explained by a state dependent stochastic model of epithelial stem cell organization. J. Theor. Biol. 186, 41–54 (1997)

9. Gerike, T., Paulus, U., Potten, C., Loeffler, M.: A dynamic model of proliferation and differentiation in the intestinal crypt based on a hypothetical intraepithelial growth factor. Cell Prolif. 31, 93–110 (1998)

10. Regev, A., Shapiro, E.: Cellular abstractions: Cells as computation. Nature 419, 343 (2002)

11. Regev, A., Shapiro, E.: The pi-calculus as an abstraction for biomolecular systems. In: Modelling in Molecular Biology. Springer, Heidelberg (2004)

12. Heath, J., Kwiatkowska, M., Norman, G., Parker, D., Tymchyshyn, O.: Probabilistic model checking of complex biological pathways. Theor. Comput. Sci. 319, 239–257 (2007)

13. Gillespie, D.: A general method for numerically simulating the stochastic time evolution of coupled chemical reactions. J. Comp. Phys. 22, 403–434 (1976)

14. Logan, C.Y., Nusse, R.: The Wnt signaling pathway in development and disease. Annu. Rev. Cell Dev. Biol. 20, 781–810 (2004)

15. Lee, E., Salic, A., Kruger, R., Heinrich, R., Kirschner, M.W.: The roles of APC and Axin derived from experimental and theoretical analysis of the Wnt pathway. PLoS Biol. 1, 10 (2003)

16. Leeuwen, I., Byrne, H., Jensen, O., King, J.: Elucidating the interactions between the adhesive and transcriptional functions of beta-catenin in normal and cancerous cells. J. Theor. Biol. 247, 77–102 (2007)

17. Regev, A., Panina, E.M., Silverman, W., Cardelli, L., Shapiro, E.: BioAmbients: an abstraction for biological compartments. Theor. Comput. Sci. 325, 141–167 (2004)

18. Potten, C.S., Kellett, M., Rew, D.A., Roberts, S.A.: Proliferation in human gastrointestinal epithelium using bromodeoxyuridine in vivo: data for different sites, proximity to a tumour, and polyposis coli. Gut 33, 524–529 (1992)

19. Cardelli, L., Gordon, A.: Mobile ambients. Theor. Comput. Sci. 240, 177–213 (2006)

20. Brittan, M., Wright, N.A.: The gastrointestinal stem cells. Cell Prolif. 37, 35–53 (2004)

21. He, X.C., Yin, T., Grindley, J.C., Tian, Q., Sato, T., Tao, W.A., Dirisina, R., Porter-Westpfahl, K.S., Hembree, M., Johnson, T., Wiedemann, L.M., Barrett, T.A., Hood, L., Wu, H., Li, L.: PTEN-deficient intestinal stem cells initiate intestinal polyposis. Nat. Genet. 39, 189–198 (2007)

22. Wong, W.M., Mandir, N., Goodlad, R.A., Wong, B.C., Garcia, S.B., Lam, S.K., Wright, N.A.: Histogenesis of human colorectal adenomas and hyperplastic polyps: the role of cell proliferation and crypt fission. Gut 50, 212–217 (2002)

23. Sansom, O.J., Reed, K.R., Hayes, A.J., Ireland, H., Brinkmann, H., Newton, I.P., Batlle, E., Simon-Assmann, P., Clevers, H., Nathke, I.S., Clarke, A.R., Winton, D.J.: Loss of Apc in vivo immediately perturbs Wnt signaling, differentiation, and migration. Genes Dev. 18, 1385–1390 (2004)
24. Gregorieff, A., Pinto, D., Begthel, H., Destree, O., Kielman, M., Clevers, H.: Expression pattern of Wnt signaling components in the adult intestine. Gastroenterology 129, 626–638 (2005)
25. http://www.cs.bham.ac.uk/~oxt/fmsb/cell.html
26. Tymchyshyn, O.: On the use of process algebra techniques in computational modelling of cancer initiation and development. PhD Thesis, University of Birmingham (forthcoming)

Approximating Continuous Systems by Timed Automata[*]

Oded Maler[1] and Grégory Batt[2]

[1] Verimag-UJF-CNRS, 2 Av. de Vignate, 38610 Gières, France
Oded.Maler@imag.fr
[2] INRIA Rocquencourt, 78153 Le Chesnay, France
Gregory.Batt@inria.fr

Abstract. In this work we develop a new technique for over-approximating (in the sense of timed trace inclusion) continuous dynamical systems by timed automata. This technique refines commonly-used discrete abstractions which are often too coarse to be useful. The essence of our technique is the partition of the state space into cubes and the allocation of *a clock for each dimension*. This allows us to get much better approximations of the behavior. We specialize this technique to multi-affine systems, a class of nonlinear systems of primary importance for the analysis of biochemical systems and demonstrate its applicability on an example taken from synthetic biology.

1 Introduction

Rigorous reasoning about the behavior of continuous dynamical systems has been a topic of study within various communities including qualitative physics in AI, robotics, and hybrid control systems. A more recent motivation comes from the domain of *systems biology* which, among other things, attempts to build *quantitative dynamic models* that capture the behavior of complex networks involving a large number of biochemical substances. Due to experimental limitations, such models admit a lot of uncertainty concerning parameter values and environmental conditions. Consequently, there is a lot of ongoing effort to apply methodologies used in the design of complex artificial systems, formal verification included, to analyze the implication of proposed models and assess their plausibility. The fact that biochemical models are often described as differential equations, with state variables denoting substance *concentrations* motivates the effort to adapt algorithmic verification technology (model checking) to continuous and hybrid systems in order to prove satisfaction of temporal properties by all system behaviors (trajectories) departing from a possible set of initial state and subject to a class of admissible inputs (disturbances).

One can classify various approaches to algorithmic verification of continuous and hybrid systems as using *direct* and *indirect* methods:[1]

---

[*] This work was partially supported by the French-Israeli project *Computational Modeling of Incomplete Biological Regulatory Networks*.

[1] Another way to view this classification is between methods based on time and space discretization.

J. Fisher (Ed.): FMSB 2008, LNBI 5054, pp. 77–89, 2008.

1. Direct methods work on the original dynamics of the system, starting from a set of initial states and applying a "successor" operator that computes the set of states reachable from those by following the continuous dynamics, until a fixpoint is reached (or not). For hybrid systems with a very simple continuous dynamics in each discrete state, namely, constant derivatives as in timed automata or linear hybrid automata (LHA), such successor states can be computed exactly for all future time instants [ACH+95, HHW97, F05]. The problem however still remains undecidable for most interesting classes due to the combination of such dynamics with discrete transitions [HKPV98, AMP95]. If the system admits a more complex dynamics defined by differential equations, the successors can be computed in an approximate manner using a kind of set-based numerical integration [DM98, CK98], [ABDM00, CK03, ADF+06].

2. Indirect methods (which are the subject of this paper) transform the original system model into an abstract model belonging to a simpler class, whose verification is easier and often decidable. The most commonly-used class of abstract models are, of course, finite-state automata, used extensively in the biological context [dJPHG01, BRdJ+05, HKI+07], but other reduction techniques have been proposed such as using timed automata to approximate continuous systems [SKE00] and LHA [OSY94], approximating continuous systems by LHA [HHW98, F05] or approximating nonlinear systems by piecewise-affine differential equations [ADG03]. The major advantage of the indirect approach is that simpler classes of models, for example finite-state automata, admit well-known model-checking algorithms, realized by numerous mature tools, while the adaptation of such techniques to systems with non-trivial continuous dynamics is much more difficult if not impossible.

Procedures for deriving such abstract models offer a tradeoff between the accuracy of the obtained model and the difficulty in deriving and analyzing it. The most straightforward approach for constructing automata from continuous systems, defined via an equation of the form $\dot{x} = f(x)$ consists of partitioning the continuous state space into rectangular cells, and defining a transition between neighboring cells if there is a trajectory of the continuous system that goes directly from one cell to another. This latter fact can be determined *locally* by evaluating f on their common boundary. While this approach guarantees a conservative over-approximation in the sense that the existence of a trajectory from x to x' in the concrete systems implies the existence of a corresponding run in the automaton, it suffers, like any abstraction technique, from "false transitivity" leading to numerous spurious behaviors, that is, abstract behaviors that do not correspond to concrete ones.

In this paper we refine this abstraction scheme by adding clocks to the automata [AD94]. The use of timed automata brings the following advantages:

1. The added clocks keep track of the progress of the trajectories along each dimension, and their values are used to constraint the dynamics of the automaton, resulting in a significant reduction of false transitivity. Moreover, the accuracy of the model can be improved indefinitely by refining the underlying grid;

2. The timed model generates timed behaviors that can be checked against *quantitative* timing properties expressed in real-time temporal logics such as MTL [Koy90] or MITL [AFH96], while this information is absent from purely discrete models;

3. The constructed models can be handled by existing verification tools for timed automata such as Uppaal [LPY97] or IF [BGM02] that can compute reachable states and, in principle, perform model checking.

The rest of the paper is organized as follows. In Section 2 we give preliminary definitions and demonstrate the problem of false transitivity. In Section 3 we show how to derive a timed automaton from a continuous dynamical system and prove that it constitutes a conservative approximation. We then present the derivation of delay bounds for the class of multi-affine systems, a class of nonlinear dynamical system used extensively in biological modeling. Section 5 reports preliminary experimental results using a prototype implementation which generates timed automata written in the IF format. A discussion of past and future work concludes the paper.

2 Preliminaries

We start this section with some definitions concerning dynamical systems, the partition of space into cubes and related geometrical concepts and notations taken from [BMP99]. To simplify notations we consider integer grids and temporal properties generated from atomic propositions of the form $x_i \geq k$ with integer k. Of course, all the results can be adapted to non-uniform grids.

We consider a dynamical system $S = (X, f)$ with state space

$$X = X_1 \times \cdots \times X_n = [0, m) \times \cdots \times [0, m) \subseteq \mathbb{R}^n$$

and dynamics is defined by

$$\dot{x} = f(x) \tag{1}$$

where $f = (f^1, \ldots, f^n)$ is a well-behaving continuous function from $\mathbb{R}^n$ to itself. A trajectory of the system starting from an initial state x is a function $\xi : \mathbb{R}_{\geq 0} \to X$ such that ξ is the solution of (1) with initial condition x_0, that is, $\xi(0) = x_0$ and for every $t \geq 0$,

$$\frac{d\xi}{dt}(t) = f(\xi(t)).$$

We impose an integer grid on X by letting $V = V_1 \times \cdots \times V_n$, $V_i = \{0, \ldots, m-1\}$ and letting $\mathcal{C}(X)$ be the set of unit cubes with integer vertices which are contained in X. We use V to represent $\mathcal{C}(X)$.

Definition 1 (Cubes, Neighbors, Facets and Slices)

1. *The cube associated with a point $v = (v_1, \ldots, v_n) \in V$ is*

$$X_v = [v_1, v_1 + 1) \times \cdots \times [v_n, v_n + 1),$$

   that is, the unit cube for which v is the leftmost corner.
2. *The successor and predecessor of a vertex/cube v in the ith direction are, respectively*

$$\sigma^{+i}(v_1, \ldots, v_{i-1}, v_i, \ldots, v_n) = (v_1, \ldots, v_{i-1}, v_i + 1, \ldots, v_n)$$

and

$$\sigma^{-i}(v_1, \ldots, v_{i-1}, v_i, \ldots, v_n) = (v_1, \ldots, v_{i-1}, v_i - 1, \ldots, v_n).$$

Two cubes/vertices are neighbors if one is the i-successor/predecessor of the other;

3. *The common facet between two neighboring cubes is the $(n-1)$-dimensional cube obtained by intersecting their boundaries;*

4. *The i-slice associated with an integer r is the set $X_{i,r}$ obtained by restricting X to points satisfying $r \leq x_i < r + 1$.*

Note that a unit cube X_v, $v = (v_1, \ldots, v_n)$, is an intersection of n slices:

$$X_v = \bigcap_{i=1}^{n} X_{i,v_i}.$$

Definition 2 (Grid Based Abstraction)

1. *The abstraction function $\alpha : X \to V$ maps every point to the cube it belongs to, that is, $\alpha(x) = v$ if $x \in X_v$;*

2. *The timed abstraction of a trajectory ξ is $\xi' = \alpha(\xi)$ such that for every t, $\xi'(t) = \alpha(\xi(t))$;*

3. *The untimed abstraction $\bar{\alpha}(\xi)$ of ξ is the (stutter-free) sequence of cubes appearing in $\alpha(\xi)$.*

Definition 3 (Extremal Values of f)

1. *The extremal values of f_i in a cube v are*

$$\underline{f}_v^i = \min\{f_i(x) : x \in X_v\} \quad and \quad \overline{f}_v^i = \max\{f_i(x) : x \in X_v\}.$$

2. *The minimal absolute velocity of f_i in a cube v is*

$$\underline{\underline{f}}_v^i = \min\{|f_i(x)| : x \in X_v\}$$

3. *The extremal values of f_i on slice $X_{i,r}$ are*

$$\underline{f}_{i,r} = \min\{f_i(x) : x \in X_{i,r}\} \quad and \quad \overline{f}_{i,r} = \max\{f_i(x) : x \in X_{i,r}\}$$

The standard way to derive a finite-state automaton from a dynamical system is summarized by the following definition.

Definition 4 (Abstraction by Automata). *The automaton $\bar{\mathcal{A}} = (V, \bar{\delta})$ is an abstraction of S if $\bar{\delta}$ consists of all pairs $(v, \sigma^{+i}(v))$ of cubes such that f_i admits a positive value on their common facet and all pairs $(v, \sigma^-(v))$ such that f_i admits a negative value on their common facet.*

Claim (Conservativism). For every trajectory ξ of S, there is a run $\bar{\xi}$ of $\bar{\mathcal{A}}_S$ such that $\bar{\xi} = \bar{\alpha}(\xi)$.

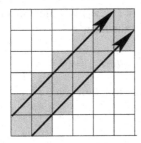 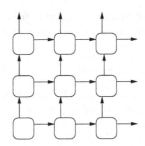

Fig. 1. (a): A simple continuous system with constant derivatives. The states reachable from the initial cube lie between the two arrows and their cube abstraction is shaded; (b) The automaton derived according to Definition 4 in which the whole state space is reachable.

This result implies that any (next-free) LTL property, generated by atoms of the form $x_i \geq k$, which is satisfied by $\bar{\mathcal{A}}_S$ is satisfied by S. However, as the following example shows, $\bar{\mathcal{A}}_S$ may have so many spurious behaviors, that it might be hard to prove interesting properties based on it. Consider a system where $f = (1, 1, \ldots, 1)$ as in Figure 1-(a). Since f has a positive component for every direction everywhere, there will be a transition from each cube to each of its i-successors and the whole state space will be reachable.

As one can see, the false transitivity is due to the fact that the transition relation between neighboring cubes is computed *locally*: since it is possible to go from v to $\sigma^{+i}(v)$ and from $\sigma^{+i}(v)$ to $\sigma^{+i}(\sigma^{+i}(v))$, the automaton allows these two successive transitions to happen, ignoring timing constraints related to the fact that between these two transitions, the trajectory needs to *cross the distance* between v_i and $v_i + 1$ in direction i, a process that takes time and might be slower then the crossing in other directions. In this paper we use clocks to impose such timing constraints.

Definition 5 (Timed Automaton). *A timed automaton is a tuple* $\mathcal{A} = (Q, \mathcal{C}, I, \Delta)$ *where Q is a finite set of discrete states, $\mathcal{C}$ is a set of clock variables ranging over* $\mathbb{R}_{\geq 0} \cup \{\perp\}$ *where $\perp$ is a special symbol indicating that the clock is inactive, I is the staying condition (invariant) which assigns to every state q, a conjunction I_q of conditions of the form $c < d$ for clock c and integer d. The transition relation Δ consists of tuples of the form (q, g, ρ, q') where q and q' are discrete states, the transition guard g is a positive combination of conditions of the form $c \geq d$ or $c = \perp$, and ρ is a clock transformation defined by one or more assignments of the form $c := 0$ or $c := \perp$.*

A *configuration* of the automaton is a pair $(q, \mathbf{z})$ where $\mathbf{z}$ is a clock valuation. The behavior of a timed automaton consists of an alternation between *time progress* periods where the automaton stays in a state q and I_q continuously holds, and *discrete instantaneous transitions* guarded by clock conditions. Formally, a *step* of the automaton is one of the following:

- A time step: $(q, \mathbf{z}) \xrightarrow{t} (q, \mathbf{z} + t)$, $t \in \mathbb{R}_+$ such that $\mathbf{z} + t$ satisfies I_q, and $\mathbf{z} + t$ is the result of adding t to clocks active in $\mathbf{z}$.
- A discrete step: $(q, \mathbf{z}) \xrightarrow{\delta} (q', \mathbf{z}')$, for some transition $\delta = (q, g, \rho, q') \in \Delta$, such that $\mathbf{z}$ satisfies g and $\mathbf{z}'$ is the result of applying ρ to $\mathbf{z}$

A *run* of the automaton starting from a configuration $(q_0, \mathbf{z}_0)$ is a finite or infinite sequence of alternating time steps and discrete steps of the form

$$\xi: \quad (q_0, \mathbf{z}_0) \xrightarrow{t_1} (q_0, \mathbf{z}_0 + t_1) \xrightarrow{\delta_1} (q_1, \mathbf{z}_1) \longrightarrow \cdots$$

whose *duration* is $\sum t_i$. One can also view such a run as a function $\xi: \mathbb{R}_{\geq 0} \to Q$ with $\xi(t) = q$ if after a duration of t the run is at state[2] q.

3   From Dynamical Systems to Timed Automata

We first establish some upper bounds on the time a trajectory may stay in a cube or in a slice and lower bounds on the time that must elapse between two successive transitions in the same direction.

Claim (Cube Sojourn Bounds). A trajectory entering a cube X_v cannot stay there more that $\bar{t}_v$ time where

$$\bar{t}_v = \min\{1/\underline{f}_v^i : 1 \leq i \leq n\}.$$

This implies that any cube X_v must be left in finite time unless every f_i attains a zero in it. The following definition establishes lower bounds on the time is takes a trajectory to cross a unit of distance in a positive or a negative direction based on the bounds on its derivative.

Claim (Slice Sojourn Bounds). Let ξ be a one-dimensional trajectory whose derivative in the interval $[t, t+h]$ is bounded in $[\underline{f}, \overline{f}]$. Then

$$\xi(t+h) - \xi(t) = 1 \Rightarrow h \geq \underline{t}^+$$
$$\xi(t+h) - \xi(t) = -1 \Rightarrow h \geq \underline{t}^-$$
$$(\forall h' \leq h \; \xi(t+h') - \xi(t) \geq -1) \Rightarrow h \leq \overline{t}^+$$
$$(\forall h' \leq h \; \xi(t+h') - \xi(t) \leq -1) \Rightarrow h \leq \overline{t}^-$$

where $\underline{t}^+, \overline{t}^+, \underline{t}^-$ and $\overline{t}^-$ are computed from $[\underline{f}, \overline{f}]$ according to the following table

$$
\begin{array}{|c||c|c||c|c|}
\hline
 & \underline{t}^+ & \overline{t}^+ & \underline{t}^- & \overline{t}^- \\
\hline
0 < \underline{f} < \overline{f} & 1/\overline{f} & 1/\underline{f} & \infty & \infty \\
\underline{f} < \overline{f} < 0 & \infty & \infty & -1/\underline{f} & -1/\overline{f} \\
\underline{f} < 0 < \overline{f} & 1/\overline{f} & \infty & -1/\underline{f} & \infty \\
\hline
\end{array}
\tag{2}
$$

Corollary 1 (Slice Transversal). *Let $\underline{f}_{i,r}$ and $\overline{f}_{i,r}$ be the bounds for f_i in slice $X_{i,r}$ and let $\underline{t}^+_{i,r}, \underline{t}^-_{i,r}, \overline{t}^+_{i,r}$ and $\overline{t}^-_{i,r}$ be the sojourn bounds derived from them according to (2).*

1. *A trajectory that enters $X_{i,r}$ from the left cannot leave it from the right in time smaller then $\underline{t}^+_{i,r}$ and cannot stay in the slice more than $\overline{t}^+_{i,r}$*

---

[2] If one or more transitions occur at t we take $\xi(t)$ to be the state reached after the last transition.

2. *A trajectory that enters $X_{i,r}$ from the right cannot leave it from the left in time smaller then $\underline{t}_{i,r}^-$ and cannot stay in the slice more than $\overline{t}_{i,r}^-$.*

Based on this bounds we can now define the approximating timed automaton. Clocks z_i^+ and z_i^- will be reset upon entering an i-slice from the left or from the right, respectively and will constrain further transitions in the same direction. Clock z will be reset at every transition and will be used for the invariant. We used timed automata with explicit deactivation of clocks denoted by $x := \perp$. Whenever a transition in one direction is taken, the clock in the other direction becomes inactive.

Definition 6 (Approximating TA).
Given a dynamical systems $S = (X, f)$, its approximating timed automaton is $\mathcal{A}_S = (V, Z, I, \Delta)$ where $Z = \{z, z_1^+, \ldots, z_n^+, z_1^-, \ldots, z_n^-\}$ is a set of clocks, I is an invariant defined for every state v as

$$I_v = z < \overline{t}_v \wedge \bigwedge_{i=1}^{n} (z_i^+ < \overline{t}_{i,v_i}^+) \wedge (z_i^- < \overline{t}_{i,v_i}^-)$$

with $z < \infty$ interpreted as true. *The transition relation Δ consists of the following transition types:*

$$\delta_v^{+i} : (v, z_i^+ \geq \underline{t}_{i,v_i}^+ \vee z_i^+ = \perp, z_i^+ := 0; z_i^- := \perp; z := 0, \sigma^{+i}(v))$$

and

$$\delta_v^{-i} : (v, z_i^- \geq \underline{t}_{i,v_i}^- \vee z_i^- = \perp, z_i^- := 0; z_i^+ := \perp; z := 0, \sigma^{-i}(v))$$

provided that such transitions are possible in the discrete abstraction $\overline{A}$.

We do not specify the initial state to provide for queries concerning different initial cubes. For every cube X_v we will use

$$Z_v = \{0\} \times [0, \underline{t}_{1,v_1}^+] \times \cdots \times [0, \underline{t}_{n,v_n}^+] \times [0, \underline{t}_{1,v_1}^-] \times \cdots \times [0, \underline{t}_{n,v_n}^-]$$

as an initial timed zone when we ask queries about trajectories starting at X_v. This way we are conservative with respect to all possible initial points in X_v which can be as close as we want to the boundary and cross as early as we want. The property of the timed automaton is summarized by the following theorem.

Theorem 1 (Neo Conservatism). *For every trajectory ξ of S starting from a point $x \in X_v$ there is at least one run ξ' of $\mathcal{A}_S$ staring from (v, Z_v) such that $\xi' = \alpha(\xi)$.*

Proof. Note that $\xi' = \alpha(\xi)$ means that ξ' takes transitions exactly when ξ crosses grid boundaries, and that ξ' can stay in a state as long as ξ stays in a cube. We use the following auxiliary assertion that we prove by induction: for every trajectory ξ of duration t, starting from $x = (x_1, \ldots, x_n) \in X_v$ and ending in $x' = (x_1', \ldots, x_n') \in X_{v'}$ there is a run ξ' of $\mathcal{A}_S$ starting at (v, Z_v) and ending in $(v', \mathbf{z})$ such that $\xi' = \alpha(\xi)$ and

1. The value of z is the time elapsed having entered X_v (or since time zero if we are still in the initial cube).

2. The value of each clock z_i^+ and z_i^- is equal to one of the following
   - The time elapsed since time zero if no crossing in direction i has occurred in ξ up to time t;
   - The time elapsed since the last i-crossing if it took place in the matching direction (positive for z_i^+, negative for z_i^-);
   - Otherwise, if the last i-crossing was in the opposite direction, the clock is inactive.

The proof is by induction on the number k of boundary crossings by the trajectory during the interval $[0, t]$:

- Base case: $k = 0$ and the trajectory remains in the same initial cube (Figure 2-(a)). According to Rolle's theorem each f_i has a derivative $(x_i' - x_i)/t$ in the cube v (and in each of the slices it belongs to). In other words

$$\underline{f}_v^i \leq (x_i' - x_i)/t \leq \overline{f}_v^i \quad \text{and} \quad \underline{f}_v^i \leq |(x_i' - x_i)|/t.$$

Taking into account that $x_i' - x_i \leq 1$ we have

$$t \leq t/(x_i' - x_i) \leq \overline{t}_{i,v_i} \quad \text{and} \quad t \leq t/|(x_i' - x_i)|/t \leq \overline{t}_v$$

which implies that there is a run of the automaton starting at v with all clocks set to zero that will satisfy the state invariants during the whole interval $[0, t]$.

- Inductive case: assuming the claim holds for all trajectories that cross grid boundaries at most k times, we show it holds for trajectories with $k + 1$ crossings. Let the new crossing occur in dimension i and, without loss of generality, be in the positive direction. First we show that for each j the state invariant holds between the last j-crossing and the time it reached x'. Since this part of the trajectory involves a displacement of length smaller than 1 in dimension j, a reasoning similar to the base case applies (Figure 2-(b)). Concerning direction i, we need, in addition, to show that the transition guard associated with the last crossing holds. Let t' denote the time between these two crossings that occur at y and y', respectively. There are two cases:

  1. They cross in the same direction, that is, at points $y = (y_1, \ldots, y_{i-1}, r, \ldots, y_n)$ and $y' = (y_1', \ldots, y_{i-1}', r + 1, \ldots, y_n)$ (Figure 2-(c)). According to the inductive hypothesis the value of clock z_i^+ when the trajectory reaches y' is the time elapsed since y. Since the i-distance is 1, by Rolle's theorem there is a derivative $1/t'$ in slice $X_{i,r}$ and the transition guard on z_i^+ will be satisfied.
  2. The crossing occurs in the opposite direction (Figure 2-(d)), hence clock z_i^+ is inactive and the transition is enabled.

  The other inductive conditions concerning clock values at time t are maintained by construction. ∎

The accuracy of the timed model can be further improved by tightening the timing constraints associated with each cube X_v. Rather then computing them based on the extremal values of f_i in the *whole slice* X_{i,v_i} we can restrict the optimization of f_i to those cubes on the slice from which X_v is indeed reachable. For example, one can

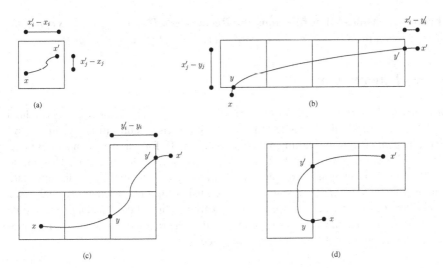

Fig. 2. (a) For a trajectory that makes no crossings, the clocks satisfy the invariant of v; (b) For a trajectory whose last crossing is in dimension i, the clocks satisfy the slice invariants associated with every j; (c) For a trajectory that crosses in direction $+i$ twice, the clocks satisfy the guard; (d) For a trajectory that crosses in dimension i in two opposite directions, the guard is trivially satisfied

observe that if for every $j \neq i$, f_j is always positive in the slice, the only cubes in the slice from which $v = (v_1, \ldots, v_i, \ldots, v_n)$ can be reached, while staying in the slice, are those of the form $v' = (v'_1, \ldots v'_i, \ldots v'_n)$ satisfying $v'_j \leq v_j$ for every j. A systematic way to obtain such restrictions is to use the untimed abstraction $\bar{A}$. Let $\pi^i(v)$ be the set of cubes v' such that there exists a run of $\bar{A}$ from v' to v which stays in X_{i,v_i}. Then we can replace the slice-based bounds $\underline{t}^+_{i,r}$, $\underline{t}^-_{i,r}$, $\overline{t}^+_{i,r}$ and $\overline{t}^-_{i,r}$ by cube-specific bounds computed according to (2), but using the extremal values of f_i on $\pi^i(v)$. Note that after deriving the timed automaton, one can apply reachability analysis on the timed automaton, obtain a subset of $\pi^i(v)$, re-compute the bound according to it and so on.

As the alert reader might have noticed we have not yet specified *how* to compute the extremal values of each f_i over cubes or slices. For linear systems, extremal values are obtained on vertices while for arbitrarily nonlinear systems one can apply numerical optimization algorithms and add some error margins to the obtained results to guarantee conservativeness. In the sequel we show how the technique specializes for *multi-affine systems*, sometimes called multi-linear systems, which are based on functions whose optimization over hyperrectangles is particularly easy.

Definition 7 (Multi-Affine Functions). *A function $p : \mathbb{R}^n \to \mathbb{R}$ is multi-affine if it is a polynomial such that the maximal degree of each variable in every term is at most one. A dynamical system $S = (X, f)$ with $f = (f_1, \ldots, f_n)$ is multi-affine if each f_i is a multi-affine function.*

The following result, due to Belta and Habets [BH06], provides for simple computation of of $\underline{f}$ and $\overline{f}$ over cubes and slices.

Theorem 2 (Multi-Affine functions and Rectangles). *The extremal values of a multi-affine function* $p : \mathbb{R}^n \to \mathbb{R}$ *over a hyperrectangle are obtained on its vertices.*

4   Implementation

Our current implementation is still in a prototype stage and it major weakness is that it works *offline*, that is, it takes a description of a piecewise[3] multi-affine dynamical systems and generates from it a timed automaton in the IF format, based on an optimized version of Definition 6. This automaton is then analyzed by the IF toolset. This implies that the timed automaton is not generated on the fly and its number of discrete states is almost the size of the grid, slightly reduced using untimed reachability analysis. A tighter integration between the approximation algorithm and the reachability computation on timed automata will allow us to restrict the generation of the TA to the reachable (under timing constraints) part of the state space.

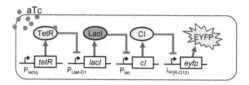

Fig. 3. The transcriptional cascade of [HTW05]

We illustrate the applicability of our approach by analyzing the timed behavior of a synthetic gene network, the cascade of transcriptional inhibitions built in *E.coli* as described in [HTW05] and illustrated in Figure 3. The cascade is made of four genes: *tetR*, *lacI*, *cI*, and *eyfp* that code, respectively for, three repressor proteins, TetR, LacI, and CI, and the fluorescent protein EYFP. The fluorescence of the system, due to the protein EYFP, is the measured output. The system can be controlled by the addition or removal of a small diffusible molecule aTc that binds to TetR and relieves the repression of *lacI* in the growth media. The transient and steady-state behavior of the system was experimentally compared with that of similar, shorter cascades [HTW05]. It was found that longer cascades have a more pronounced ultrasensitive input/output responses at steady-state, but longer response times. A modifications of biological parameters that should improve the ultrasensitive response was proposed in [BYWB07], but the potential modifications of network *response times* has not been investigated.

To investigate this question we cut the 5-dimensional state space into more than 2000 cubes, from which we generate a timed automaton. We then use IF to check whether a significant increase of the fluorescence of the system (from less than 500 to more than 5000 fluorescence units) is obtained in a reasonable time following the addition of aTc in the growth medium. For the original system, prior to the improvement proposed in [HTW05], our analysis shows that the required increase is guaranteed to happen in at

---

[3] The extension of our results to *piecewise* multi-affine systems which are continuous on the switching boundaries is straightforward.

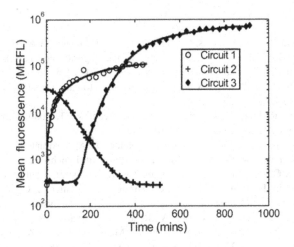

Fig. 4. Experimentally observed delay (circuit 3)

most 2820 minutes. On the other hand, for the tuned system, we obtain a significantly smaller upper bound (1680 minutes, 40% less). This suggests that the proposed modification improves the response time of the system, in addition to improving its steady state behavior. We compare these (worst case) time bounds with observations on the actual system (Figure 4) where the fluorescence of the system reaches the target value 5000 approximatively 200 minutes after addition of aTc. This clearly reveals the conservativeness of our approach, an issue that can be addressed by using finer partitions of the state space.

5 Discussion

We have developed a technique for approximating dynamical systems by timed automata for the purpose of checking timed properties. The essence of this technique, is the use of dimension-specific clocks, in contrast with the approach of [SKE00] which uses one clock (our z) for the whole cube. These ideas are close in spirit to the rectangular hybrid automata of [HKPV98], in the sense of separating and bounding the dynamics of each dimension. In that work, the emphasis was, however, on exact decidability which required a reset (initialization) of all continuous variables when a boundary is crossed, a feature which is not useful in the continuous context.

Our approach performs reasonably well in those parts of the state space where all variables admit a monotone dynamics and the major challenge is to improve its treatment of those parts of the state space where one or more derivative changes its sign, and hence admits a zero. In that case, the non existence of a finite upper bound can make it hard to prove eventuality properties. We also explore that extension of these techniques to even richer dynamics.

References

[ACH+95]   Alur, R., Courcoubetis, C., Halbwachs, N., Henzinger, T.A., Ho, P.-H., Nicollin, X., Olivero, A., Sifakis, J., Yovine, S.: The Algorithmic Analysis of Hybrid Systems. Theoretical Computer Science 138, 3–34 (1995)

[AD94]   Alur, R., Dill, D.L.: A Theory of Timed Automata. Theoretical Computer Science 126, 183–235 (1994)

[AFH96]   Alur, R., Feder, T., Henzinger, T.A.: The Benefits of Relaxing Punctuality. Journal of the ACM 43, 116–146 (1996)

[ABDM00]   Asarin, A., Bournez, O., Dang, T., Maler, O.: Approximate Reachability Analysis of Piecewise-linear Dynamical Systems. In: Lynch, N.A., Krogh, B.H. (eds.) HSCC 2000. LNCS, vol. 1790, pp. 20–31. Springer, Heidelberg (2000)

[ADF+06]   Asarin, E., Dang, T., Frehse, G., Girard, A., Le Guernic, C., Maler, O.: Recent Progress in Continuous and Hybrid Reachability Analysis. In: CACSD (2006)

[ADG03]   Asarin, E., Dang, T., Girard, A.: Reachability Analysis of Nonlinear Systems using Conservative Approximation. In: Maler, O., Pnueli, A. (eds.) HSCC 2003. LNCS, vol. 2623, pp. 20–35. Springer, Heidelberg (2003)

[AMP95]   Asarin, E., Maler, O., Pnueli, A.: Reachability Analysis of Dynamical Systems having Piecewise-Constant Derivatives. Theoretical Computer Science 138, 35–65 (1995)

[BRdJ+05]   Batt, G., Ropers, D., de Jong, H., Geiselmann, J., Mateescu, R., Page, M., Schneider, D.: Validation of qualitative models of genetic regulatory networks by model checking: Analysis of the nutritional stress response in *Escherichia coli*. Bioinformatics 21, i19–i28 (2005)

[BYWB07]   Batt, G., Yordanov, B., Weiss, R., Belta, C.: Robustness analysis and tuning of synthetic gene networks. Bioinformatics 23, 2415–2422 (2007)

[BH06]   Belta, C., Habets, L.C.G.J.M.: Controlling a Class of Nonlinear Systems on Rectangles. IEEE Trans. on Automatic Control 51, 1749–1759 (2006)

[BT00]   Botchkarev, O., Tripakis, S.: Verification of Hybrid Systems with Linear Differential Inclusions using Ellipsoidal Approximations. In: Lynch, N.A., Krogh, B.H. (eds.) HSCC 2000. LNCS, vol. 1790, pp. 73–88. Springer, Heidelberg (2000)

[BMP99]   Bournez, O., Maler, O., Pnueli, A.: Orthogonal Polyhedra: Representation and Computation. In: Vaandrager, F.W., van Schuppen, J.H. (eds.) HSCC 1999. LNCS, vol. 1569, pp. 46–60. Springer, Heidelberg (1999)

[BGM02]   Bozga, M., Graf, S., Mounier, L.: IF-2.0: A Validation Environment for Component-Based Real-Time Systems. In: Brinksma, E., Larsen, K.G. (eds.) CAV 2002. LNCS, vol. 2404, pp. 343–348. Springer, Heidelberg (2002)

[CK98]   Chutinan, A., Krogh, B.H.: Computing Polyhedral Approximations to Dynamic Flow Pipes. In: CDC 1998. IEEE, Los Alamitos (1998)

[CK03]   Chutinan, A., Krogh, B.H.: Computational Techniques for Hybrid System Verification. IEEE Trans. on Automatic Control 8, 64–75 (2003)

[DM98]   Dang, T., Maler, O.: Reachability Analysis via Face Lifting. In: Henzinger, T.A., Sastry, S.S. (eds.) HSCC 1998. LNCS, vol. 1386, pp. 96–109. Springer, Heidelberg (1998)

[dJPHG01]   de Jong, H., Page, M., Hernandez, C., Geiselmann, J.: Qualitative Simulation of Genetic Regulatory Networks: Method and Application. In: IJCAI-2001, pp. 67–73 (2001)

[F05]   Frehse, G.: PHAVer: Algorithmic Verification of Hybrid Systems Past HyTech. In: Morari, M., Thiele, L. (eds.) HSCC 2005. LNCS, vol. 3414, pp. 258–273. Springer, Heidelberg (2005)

[HKI+07]   Halasz, A., Kumar, V., Imielinski, M., Belta, C., Sokolsky, O., Pathak, S.: Analysis of Lactose Metabolism in *E.coli* using Reachability Analysis of Hybrid Systems. IEE Proceedings - Systems Biology 21, 130–148 (2007)
[HHW98]   Henzinger, T.A., Ho, P.-H., Wong-Toi, H.: Algorithmic Analysis of Nonlinear Hybrid Systems. IEEE Trans. on Automatic Control 43, 540–554 (1998)
[HKPV98]   Henzinger, T.A., Kopke, P.W., Puri, A., Varaiya, P.: What's Decidable about Hybrid Automata? Journal of Computer and System Sciences 57, 94–124 (1998)
[HTW05]   Hooshangi, S., Thiberge, S., Weiss, R.: Ultrasensitivity and noise propagation in a synthetic transcriptional cascade. PNAS 102, 3581–3586 (2005)
[HHW97]   Henzinger, T.A., Ho, P.-H., Wong-Toi, H.: Hytech: A Model Checker for Hybrid Systems. Software Tools for Technology Transfer 1, 110–122 (1997)
[Koy90]   Koymans, R.: Specifying Real-time Properties with Metric Temporal Logic. Real-time Systems 2, 255–299 (1990)
[LPY97]   Larsen, K.G., Pettersson, P., Yi, W.: Uppaal in a Nutshell. Software Tools for Technology Transfer 1, 134–152 (1997)
[OSY94]   Olivero, A., Sifakis, J., Yovine, S.: Using abstractions for the verification of linear hybrid systems. In: Dill, D.L. (ed.) CAV 1994. LNCS, vol. 818, pp. 81–94. Springer, Heidelberg (1994)
[SKE00]   Stursberg, O., Kowalewski, S., Engell, S.: On the Generation of Timed Discrete Approximations for Continuous Systems. Mathematical and Computer Models of Dynamical Systems 6, 51–70 (2000)

From Reaction Models to
Influence Graphs and Back: A Theorem[*]

François Fages and Sylvain Soliman

Projet Contraintes, INRIA Rocquencourt,
BP105, 78153 Le Chesnay Cedex, France
http://contraintes.inria.fr

Abstract. Biologists use diagrams to represent interactions between molecular species, and on the computer, diagrammatic notations are also more and more employed in interactive maps. These diagrams are fundamentally of two types: reaction graphs and activation/inhibition graphs. In this paper, we study the formal relationship between these graphs. We consider systems of biochemical reactions with kinetic expressions, as written in the Systems Biology Markup Language SBML, and interpreted by a system of Ordinary Differential Equations over molecular concentrations. We show that under a general condition of increasing monotonicity of the kinetic expressions, and in absence of both activation and inhibition effects between a pair of molecules, the influence graph inferred from the stoichiometric coefficients of the reactions is equal to the one defined by the signs of the coefficients of the Jacobian matrix. Under these conditions, satisfied by mass action law, Michaelis-Menten and Hill kinetics, the influence graph is thus independent of the precise kinetic expressions, and is computable in linear time in the number of reactions. We apply these results to Kohn's map of the mammalian cell cycle and to the MAPK signalling cascade. Then we propose a syntax for denoting antagonists in reaction rules and generalize our results to this setting.

1   Introduction

Biologists use diagrams to represent interactions between molecular species, and diagrammatic notations like the ones introduced by Kohn in his map of the mammalian cell cycle [2] are also employed on the computer in interactive maps, like for instance MIM[1]. This type of notation encompasses two types of information : interactions (binding, complexation, protein modification, etc.) and regulations (of an interaction or of a transcription).

The Systems Biology Markup Language (SBML) [3] uses a syntax of reaction rules with kinetic expressions to define reaction models in a precise way, and more and more models are described in such a formalism, like in the biomodels.net

---

[*] This paper provides a direct presentation and a generalization of one theorem shown in [1] among other results in the framework of abstract interpretation which is not used here.

[1] http://discover.nci.nih.gov/mim/

J. Fisher (Ed.): FMSB 2008, LNBI 5054, pp. 90–102, 2008.
© Springer-Verlag Berlin Heidelberg 2008

repository. This type of language is well suited to describe interactions (and in a limited manner their regulations through the notion of *modifiers*) but not directly molecule to molecule activations and inhibitions.

On the other hand, formal influence graphs for activation and inhibition have been introduced in the setting of gene regulatory networks [4] as an abstraction of complex reaction networks. These graphs completely abstract from the precise interactions, especially at post-transcriptional level, and retain only the activation and inhibition effects between genes. In these influence graphs, the existence of a positive circuit (resp. a negative circuit) has been shown to be a necessary condition for multistationarity (resp. oscillations) in different settings [5,6,7,8,9], as conjectured by Thomas [10].

There are nowadays several tools providing different kinds of analyses for either reaction models or influence graphs. However the only formal relationship relating the two seems to be the extraction of the influence graph from the Jacobian matrix derived from the reaction model, when equipped with precise kinetic expressions and parameter values.

In this paper, we study more systematically the formal relationship between reaction models and activation/inhibition influence graphs. We consider systems of biochemical reactions with kinetic expressions, as written in the Systems Biology Markup Language SBML, and interpreted by systems of Ordinary Differential Equations over molecular concentrations. We show that under the general condition of strongly increasing monotonicity of the kinetic expressions, and in absence of both activation and inhibition effects from one molecule to the same target, the influence graph inferred from the stoichiometric coefficients of the reactions, called the syntactical influence graph, is identical to the influence graph defined by the signs of the coefficients of the Jacobian matrix, called the differential influence graph. Under these conditions, satisfied by mass action law, Michaelis-Menten and Hill kinetics, the influence graph is thus independent of the kinetic expressions for the reactions, and is computable in linear time in the number of reactions.

We show that this remarkable property applies to the transcription of Kohn's map of the mammalian cell cycle control [2] into an SBML model of approx. 800 reactions [11]. On this example, the syntactical influence graph is computed in less than one second, and our equivalence theorem shows that this influence graph would be the same as the differential influence graph for any standard kinetics and any (non zero) parameter values. The same property of independence from the kinetic expressions holds for the influence graph inferred from the MAPK signalling model of Levchenko et al. [12]. This influence graph exhibits positive as well as negative feedbacks that are hidden in the purely directional cascade of the reaction graph [13], and that have been the reason for an erroneous interpretation of Thomas' rules when applied to the MAPK cascade in [14].

Finally, we consider generalized reaction rules, where inhibitors can be indicated in the syntax of the rules, and generalize our results to this setting for a large set of kinetic expressions.

2   Reaction Models

Following SBML and BIOCHAM [15,16] conventions, a model of a biochemical system is formally a set of reaction rules of the form e for S => S' where S is a set of molecules given with their stoichiometric coefficient, called a *solution*, S' is the transformed solution, and e is a kinetic expression involving the concentrations of molecules (which are not strictly required to appear in S).

We will use the BIOCHAM operators + and * to denote solutions as 2*A + B, as well as the syntax of catalyzed reactions e for S =[C]=> S' as an abbreviation for e for S+C => S'+C.

Classical kinetic expressions are the mass action law kinetics

$$k * \prod_{i=1}^{n} x_i{}^{l_i}$$

for a reaction with n reactants x_i, where l_i is the stoichiometric coefficient of x_i as a reactant, Michaelis-Menten kinetics

$$V_m * x_s/(K_m + x_s)$$

for an enzymatic reaction of the form $x_s = [x_e] => x_p$, where[2] $V_m = k * (x_e + x_e * x_s/K_m)$, and Hill's kinetics

$$V_m * x_s{}^n/(K_m^n + x_s{}^n)$$

of which Michaelis-Menten kinetics is a special case with $n = 1$.

A set of reaction rules $\{e_i \text{ for } S_i => S'_i\}_{i=1,...,n}$ over molecular concentration variables $\{x_1, ..., x_m\}$, can be interpreted under different semantics. The traditional *differential semantics* interpret the rules by the following system of Ordinary Differential Equations (ODE):

$$dx_k/dt = \sum_{i=1}^{n} r_i(x_k) * e_i - \sum_{j=1}^{n} l_j(x_k) * e_j$$

where $r_i(x_k)$ (resp. l_i) is the stoichiometric coefficient of x_k in the right (resp. left) member of rule i.

The differential semantics will be the only interpretation of reaction models considered here. In this paper, we shall not consider the other interpretations of reaction rules used in BIOCHAM [1], namely the *stochastic semantics*, where the kinetic expressions are interpreted as transition probabilities, the rule set as a continuous-time Markov chain that can be simulated with Gillespie's algorithm [17], or the *boolean semantics* which simply forgets the kinetic expressions and interpret the rules as a non-deterministic (asynchronous) transition system over boolean states representing the absence or presence of molecules.

---

[2] $x_e * x_s/K_m$ is the concentration of the enzyme-substrate complex, supposed constant in the Michaelian approximation and $x_e + x_e * x_s/K_m$ is thus the total amount of enzyme.

3   Influence Graphs of Activation and Inhibition

Influence graphs for activation and inhibition have been introduced for the analysis of gene expression in the setting of gene regulatory networks [4]. Such influence graphs are in fact an abstraction of complex reaction networks, and can be applied as such to protein interaction networks. However the distinction between the influence graph and the reaction (hyper)graph is crucial to the application of Thomas's conditions of multistationarity and oscillations [4,7] to protein interaction network, and there has been some confusion between the two kinds of graphs [14].

Here we consider two definitions of the influence graph associated to a reaction model, and show their equivalence under general assumptions.

3.1   Definition from the Jacobian Matrix

In the differential semantics of a reaction rule model $M = \{e_i \text{ for } l_i \mathrel{=\!\!>} r_i \mid i \in I\}$ we have $\dot{x}_k = dx_k/dt = \sum_{i=1}^{n}(r_i(x_k) - l_i(x_k)) * e_i$. The Jacobian matrix J is formed of the partial derivatives $J_{ij} = \partial \dot{x}_i / \partial x_j$.

Definition 1. *The* differential influence graph *associated to a reaction model is the graph having for vertices the molecular species, and for edge-set the following two kinds of edges:*

$\{A \text{ activates } B \mid \partial \dot{x}_B / \partial x_A > 0 \text{ in some point of the space}\}$
$\cup \{A \text{ inhibits } B \mid \partial \dot{x}_B / \partial x_A < 0 \text{ in some point of the space}\}$

Both activation and inhibition edges may exist between two molecular species in reaction models such as for instance:

$k_1 * A$ for $A \mathrel{=\!\!>} B$
$k_2 * A * B$ for $A + B \mathrel{=\!\!>} C$

We have indeed $dB/dt = k_1 * A - k_2 * A * B$ and $\partial \dot{B} / \partial A = k_1 - k_2 * B$, hence A *inhibits* B and A *activates* B both belong to the differential influence graph in such an example.

3.2   Definition from the Stoichiometric Coefficients

Definition 2. *The* syntactical influence graph *associated to a reaction model M is the graph having for vertices the molecular species, and for edges the following set:*

$\{A \text{ inhibits } B \quad \mid \exists (e_i \text{ for } l_i \mathrel{=\!\!>} r_i) \in M,$
$\qquad\qquad l_i(A) > 0 \text{ and } r_i(B) - l_i(B) < 0\}$
$\cup \{A \text{ activates } B \mid \exists (e_i \text{ for } l_i \mathrel{=\!\!>} r_i) \in M,$
$\qquad\qquad l_i(A) > 0 \text{ and } r_i(B) - l_i(B) > 0\}$

In particular, we have the following influences for elementary reactions of complexation, modification, synthesis and degradation:

$\alpha(\{A + B \mathrel{=\!\!>} C\}) = \{$ A inhibits B, A inhibits A, B inhibits A,
$\qquad\qquad\qquad$ B inhibits B, A activates C, B activates C$\}$
$\alpha(\{A = [C] \mathrel{=\!\!>} B\}) = \{$ C inhibits A, A inhibits A, A activates B, C activates B$\}$
$\alpha(\{A = [B] \mathrel{=\!\!>} \_\}) = \{$ B inhibits A, A inhibits A$\}$
$\alpha(\{\_ = [B] \mathrel{=\!\!>} A\}) = \{$ B activates A$\}$

The inhibition loops on the reactants are justified by the negative sign in the Jacobian matrix of the differential semantics of such reactions. Unlike the differential influence graph, this graph is clearly trivial to compute by browsing the syntax of the rules:

Proposition 1. *The syntactical influence graph of a reaction model of n rules is computable in $O(n)$ time.*

3.3   Over-Approximation Theorem

Comparing the differential influence graph and the syntactical influence graph requires that the information in the kinetic expressions and in the reactions be compatible. This motivates the following definition where the first property forbids the presence of purely kinetic inhibitors not represented in the reaction, and the second property enforces that the variables appearing in the kinetic expressions do appear as reactants or enzymes in the reaction.

Definition 3. *In a reaction rule e for l=>r, we say that a kinetic expression e is increasing iff for all molecules x_k we have*

1. *$\partial e/\partial x_k \geq 0$ in all points of the space,*
2. *$l(x_k) > 0$ if $\partial e/\partial x_k > 0$ in some point of the space.*

A reaction model has an increasing kinetics *iff all its reaction rules have an increasing kinetics.*

One can easily check that:

Proposition 2. *Mass action law kinetics for any reaction, as well as Michaelis Menten and Hill kinetics for enzymatic reactions, are increasing.*

On the other hand, negative Hill kinetics of the form $k_1/(k_2^n + y^n)$ are not increasing. They represent an inhibition by a molecule y not belonging to the reactants, and thus not reflected in the syntax of the reaction.

Theorem 1. *For any reaction model with an increasing kinetics, the differential influence graph is a subgraph of the syntactical influence graph.*

Proof. If (A activates B) belongs to the differential influence graph then $\partial \dot{B}/\partial A > 0$. Hence there exists a term in the differential equation for B, of the form $(r_i(B) - l_i(B)) * e_i$ with $\partial e_i/\partial A$ of the same sign as $r_i(B) - l_i(B)$.

Let us suppose that $r_i(B) - l_i(B) > 0$ then $\partial e_i/\partial A > 0$ and since e_i is increasing we get that $l_i(A) > 0$ and thus that (A activates B) in the syntactical graph. If on the contrary $r_i(B) - l_i(B) < 0$ then $\partial e_i/\partial A < 0$, which is not possible for an increasing kinetics.

If (A inhibits B) is in the differential graph then $\partial \dot{B}/\partial A < 0$. Hence there exists a term in the differential semantics, of the form $(r_i(B) - l_i(B)) * e_i$ with $\partial e_i/\partial A$ of sign opposite to that of $r_i(B) - l_i(B)$.

Let us suppose that $r_i(B) - l_i(B) > 0$ then $\partial e_i/\partial A < 0$, which is not possible for an increasing kinetics. If on the contrary $r_i(B) - l_i(B) < 0$ then $\partial e_i/\partial A > 0$ and since e_i is increasing we get that $l_i(A) > 0$ and thus that (A activates B) is in the syntactical influence graph.   □

Corollary 1. *For any reaction model with an increasing kinetics, the differential influence graph restricted to the phase space w.r.t. some initial conditions, is a subgraph of the syntactical influence graph.*

Proof. Restricting the points of the phase space to those points that are accessible from some initial states, restricts the number of edges in the differential influence graphs which thus remains a subgraph of the syntactical influence graph. □

It is worth noticing that even in the simple case of mass action law kinetics, the differential influence graph may be a strict subset of the syntactical influence graph. For instance let x be the following model :

$k_1 * A$ for $A => B$

$k_2 * A$ for $\_ = [A] => A$

In the syntactical influence graph, A *activates* B, A *activates* A and A *inhibits* A, however $\dot{A} = (k_2 - k_1) * A$, hence $\partial\dot{A}/\partial A$ can be made always positive or always negative or always null, resulting in the absence of respectively, A *inhibits* A, A *activates* A or both, in the differential influence graph.

3.4 Equivalence Theorem

Definition 4. *In a reaction rule e* for *l=>r, a kinetic expression e is strongly increasing iff for all molecules x_k we have*

1. $\partial e/\partial x_k \geq 0$ *in all points of the space,*
2. $l(x_k) > 0$ *if and only if there exists a point in the space s.t. $\partial e/\partial x_k > 0$*

A reaction model has a strongly increasing kinetics iff all its reaction rules have a strongly increasing kinetics.

Note that *strongly increasing* implies *increasing*.

Proposition 3. *Mass action law kinetics for any reaction, as well as Michaelis Menten and Hill kinetics for enzymatic reactions, are strongly increasing.*

Proof. For the case of Mass action law, the kinetics are of the form:

$$e_i = k_i * \prod_{l=1}^{m} x_l^{l_i(x_l)}$$

with $k_i > 0$ and $l_i(x_l) \geq 0$. We thus have $\partial e_i/\partial x_k = 0$ if $l_i(x_k) = 0$ and $\partial e_i/\partial x_k = k_i * l_i(x_k) * x_k^{l_i(x_k)-1} \prod_{l\neq k} x_l^{l_i(x_l)}$ otherwise, which clearly satisfies (1) and (2).

In the case of Hill kinetics (of which Michaelis Menten is a subcase), we have:

$$e_i = \frac{V_m * x_s^n}{K_m^n + x_s^n}$$

for the reaction $x_s + x_e => x_p + x_e$ and where $V_m = k_2 * x_e^{tot} = k_2 * (x_e + k_1 * x_e * x_s/(k_{-1} + k_2))$ from the steady state approximation. It is obvious that

$\partial e_i/\partial x_k = 0$ for all x_k other than x_s and x_e since they do not appear in e_i and one can easily check that with all the constants n, k_1, k_{-1}, k_2 strictly positive, both $\partial e_i/\partial x_e$ and $\partial e_i/\partial x_s$ are greater than 0 at some point in the space.   □

Lemma 1. *Let M be a reaction model with a strongly increasing kinetics,*

If (A activates B) is an edge in the syntactical influence graph, and not (A inhibits B), then (A activates B) belongs to the differential influence graph.

If (A inhibits B) is an edge in the syntactical influence graph, and not (A activates B), then (A inhibits B) belongs to the differential influence graph.

Proof. Since $\partial\dot{B}/\partial A = \sum_{i=1}^{n}(r_i(B) - l_i(B)) * \partial e_i/\partial A$ and all e_i are increasing we get that $\partial\dot{B}/\partial A = \sum_{\{i\leq n|l_i(A)>0\}}(r_i(B) - l_i(B)) * \partial e_i/\partial A$.

Now if (A activates B) is in the syntactical influence graph, but not (A inhibits B), then all rules such that $l_i(A) > 0$ verify $r_i(B) - l_i(B) \geq 0$ and there is at least one rule for which the inequality is strict. We thus get that $\partial\dot{B}/\partial A$ is a sum of positive numbers, amongst which one is such that $r_i(B) - l_i(B) > 0$ and $l_i(A) > 0$ which, since M is strongly increasing, implies that there exists a point in the space for which $\partial e_i/\partial A > 0$. Hence $\partial\dot{B}/\partial A > 0$ at that point, and (A activates B) is thus in the differential influence graph.

For inhibition the same reasoning applies with the opposite sign for the $r_i(B) - l_i(B)$ and thus for the partial derivative $\partial\dot{B}/\partial A$.   □

This lemma establishes the following equivalence result:

Theorem 2. *In a reaction model with a strongly increasing kinetics and where no molecule is at the same time an activator and an inhibitor of the same target molecule, the differential and syntactical influence graphs coincide.*

This theorem shows that for standard kinetic expressions, the syntactical influences coincide with the differential influences based on the signs of the coefficients in the Jacobian matrix, when no molecule is at the same time an activator and an inhibitor of the same molecule. The theorem thus provides a linear time algorithm for computing the differential influences in these cases, simply by computing the syntactical influences. It shows also that the differential influence graph is independent of the kinetic expressions.

Corollary 2. *The differential influence graph of a reaction model of n rules with a strongly increasing kinetics is computable in time $O(n)$ if no molecule is at the same time an activator and an inhibitor.*

Corollary 3. *The differential influence graph of a reaction model is independent of the kinetic expressions as long as they are strongly increasing, if no molecule is at the same time an activator and an inhibitor.*

4   Application to Kohn's Map of the Mammalian Cell Cycle Control

Kohn's map of the mammalian cell cycle control [2] has been transcribed in BIOCHAM to serve as a large benchmarking example of approx. 500 species and 800 rules [11].

On Kohn's map, the computation of activation and inhibition influences takes less than one second CPU time (on a PC 1,7GHz) for the complete model, showing the efficiency of the syntactical inference algorithm. The influence graph is composed of 1231 activation edges and 1089 inhibition edges.

Furthermore in this large example no molecule is both an activator and an inhibitor of the same target molecule. Theorem 2 thus entails that the computed influence graph is equal to the differential graph that would be obtained in any kinetic model of Kohn's map for any standard kinetic expressions and for any (non zero) parameter values.

Since there is a lot of kinetic data missing for such a big model, the possibility to nevertheless obtain the exact influence graph without having to estimate parameters or even to choose precise kinetic expressions is quite remarkable, and justifies the use of purely qualitative models for the analysis of feedback circuits.

5 Application to the Signal Transduction MAPK "cascade"

Let us consider the MAPK signalling model of [12]. Figure 1 depicts the reaction graph as a bipartite graph with round boxes for molecules and rectangular boxes for rules. Figure 2 depicts the syntactical influence graph, where activation (resp. inhibition) is materialized by plain (resp. dashed) arrows.

This computed graph reveals the negative feedbacks that are somewhat hidden in a purely directional signalling cascade of reactions. Furthermore, as no molecule is at the same time an activator and an inhibitor of a same molecule, this graph is largely independent of the kinetics of the reactions, and would be identical to the differential influence graph for any standard kinetic expressions with any (non zero) kinetic parameter values.

These negative feedbacks, a necessary condition for oscillations [4,8,9], have been formally analyzed in [13] and interpreted as enzyme sequestration in complexes. Furthermore, oscillations in the MAPK cascade model have been shown in [18].

The influence graph also exhibits positive circuits. These are a necessary condition for multistationarity [4,7] that has been observed in the MAPK model, and experimentally in Xenopus oocytes [14]. Note that the absence of circuit in the (directional) reaction graph of MAPK was misinterpreted as a counterexample to Thomas' rule in [14] because of a confusion between both kinds of graphs.

6 Adding a Syntax for Antagonists in Reaction Rules

The over-approximation theorem 1 may suggest to provide a syntax for antagonists (i.e. inhibitors) in reaction rules, and generalize the result to this setting. Note that the mixing of mechanistic reaction models with non-mechanistic

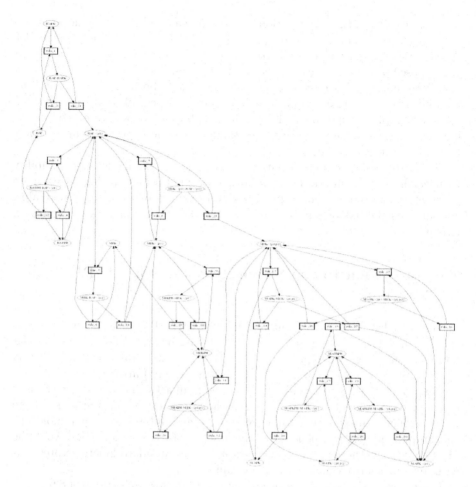

Fig. 1. Reaction graph of the MAPK model of[12]

information on the inhibitors of some reactions, is a common practice in diagrammatic notations which often combine reaction edges with activation and inhibition edges.

Let us denote by (*e* for *l* =[/*a*]=> *r*) a *generalized reaction rule* with antagonists *a*. Reaction rules with catalysts, of the form (*e* for *l* =[*c*/*a*]=> *r*), will remain an abbreviation for (*e* for *l* + *c* =[/*a*]=> *r* + *c*). This notation for antagonists thus provides a counterpart for denoting the inhibitory effect of some agent on a reaction, symmetrically to the activation effect of the catalysts of the reaction.

Definition 5. *The syntactical influence graph associated to a generalized reaction model M is the graph having for vertices the molecular species, and for edges the following set:*

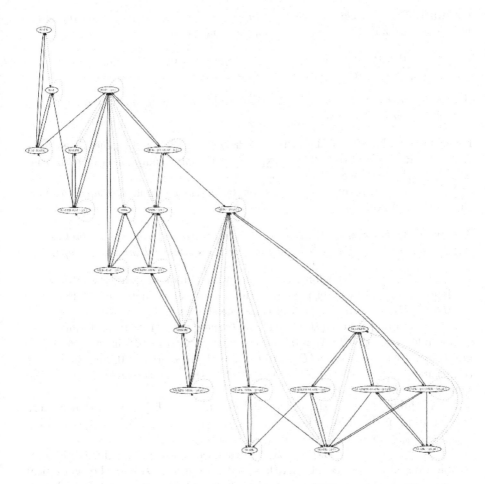

Fig. 2. Influence graph inferred from the MAPK reaction model

$$
\begin{aligned}
\{A \ \mathit{inhibits} \ B \quad &| \ \exists(e_i\texttt{for} \ l_i \ \texttt{=[/}a_i\texttt{]=>} \ r_i) \in M, \\
&\quad l_i(A) > 0 \ \mathit{and} \ r_i(B) - l_i(B) < 0\} \\
\cup\{A \ \mathit{activates} \ B \ &| \ \exists(e_i\texttt{for} \ l_i \ \texttt{=[/}a_i\texttt{]=>} \ r_i) \in M, \\
&\quad l_i(A) > 0 \ \mathit{and} \ r_i(B) - l_i(B) > 0\} \\
\cup\{A \ \mathit{activates} \ B \ &| \ \exists(e_i\texttt{for} \ l_i \ \texttt{=[/}a_i\texttt{]=>} \ r_i) \in M, \\
&\quad a_i(A) > 0 \ \mathit{and} \ r_i(B) - l_i(B) < 0\} \\
\cup\{A \ \mathit{inhibits} \ B \ \ &| \ \exists(e_i\texttt{for} \ l_i \ \texttt{=[/}a_i\texttt{]=>} \ r_i) \in M, \\
&\quad a_i(A) > 0 \ \mathit{and} \ r_i(B) - l_i(B) > 0\}
\end{aligned}
$$

For instance, the set of syntactical influences of the generalized reaction rule A =[/I]=> B} is {A *inhibits* A, I *activates* A, A *activates* B, I *inhibits* B}. On the other hand, note that the definition of the differential influence graph applies to generalized reaction models as it is based on the kinetic expressions only.

Definition 6. *In a generalized reaction rule e* for *l* =[/a]=> *r, a kinetic expression e is compatible iff for all molecules x_k we have*

1. *$l(x_k) > 0$ if there exists a point in the space s.t. $\partial e / \partial x_k > 0$,*
2. *$a(x_k) > 0$ if there exists a point in the space s.t. $\partial e / \partial x_k < 0$.*

A generalized reaction model has a compatible kinetics *iff all its reaction rules have a compatible kinetics.*

For instance, a kinetics of the form k1*Mdm2/(k2+P53) for the generalized reaction rule Mdm2 =[/P53]=> Mdm2p expressing the phosphorylation of Mdm2 that is inhibited by P53 (see [19]) is compatible.

Note that for a reaction model, *strongly increasing* implies *compatible*. Furthermore, we have:

Theorem 3. *For any generalized reaction model with a compatible kinetics, the differential influence graph is a subgraph of the syntactical influence graph.*

Proof. If (A activates B) belongs to the differential influence graph then $\partial \dot{B} / \partial A > 0$. Hence there exists a term in the differential equation for B, of the form $(r_i(B) - l_i(B)) * e_i$ with $\partial e_i / \partial A$ of the same sign as $r_i(B) - l_i(B)$.

Let us suppose that $r_i(B) - l_i(B) > 0$ then $\partial e_i / \partial A > 0$, and since e_i is compatible we get that $l_i(A) > 0$ and thus that (A activates B) in the syntactical graph. If on the contrary $r_i(B) - l_i(B) < 0$ then $\partial e_i / \partial A < 0$, and since e_i is compatible we get that $a_i(A) > 0$ and thus that (A activates B) is in the syntactical influence graph.

If (A inhibits B) is in the differential graph then $\partial \dot{B} / \partial A < 0$. Hence there exists a term in the differential semantics, of the form $(r_i(B) - l_i(B)) * e_i$ with $\partial e_i / \partial A$ of sign opposite to that of $r_i(B) - l_i(B)$.

Let us suppose that $r_i(B) - l_i(B) > 0$ then $\partial e_i / \partial A < 0$, and since e_i is compatible we get that $a_i(A) > 0$ and thus that (A inhibits B) is in the syntactical influence graph. If on the contrary $r_i(B) - l_i(B) < 0$ then $\partial e_i / \partial A > 0$, and since e_i is compatible we get that $l_i(A) > 0$ and thus that (A activates B) is in the syntactical influence graph.   □

This theorem shows that in this setting which mixes reaction rules with information on antagonists, the syntactical influence graph still over-approximates the differential influence graph for any standard kinetics.

7 Conclusion

This work shows that to a large extent, the influence graph of a reaction model is independent of the kinetic parameters and kinetic expressions, and that it can be computed in linear time simply from the syntax of the reactions. This happens for strongly increasing kinetics such as classical mass action law, Michaelis-Menten and Hill kinetics, when no molecule is at the same time an activator and an inhibitor of a same target molecule.

The inference of the syntactical influence graph from a reaction model has been implemented in BIOCHAM, and applied to various models. On a transcription of Kohn's map into approx. 800 reaction rules, this implementation shows that no molecule is at the same time an activator and an inhibitor of a same molecule, and therefore, our equivalence theorem states that the differential influence graph would be the same for any standard kinetics with any parameter values.

On the MAPK signalling cascade that does not contain any feedback reaction, the implementation does reveal both positive and negative feedback circuits in the influence graph, which has been a source of confusion for the correct application of Thomas' rules. Furthermore, in this example again, no molecule is at the same time an activator and an inhibitor of another molecule, showing the independence of the influence graph from the kinetics.

Acknowledgement. This work benefited from partial support of the European Union FP6 Network of Excellence REWERSE http://www.rewerse.net, and Strep TEMPO http://www.chrono-tempo.org.

References

1. Fages, F., Soliman, S.: Abstract interpretation and types for systems biology. Theoretical Computer Science (to appear, 2008)
2. Kohn, K.W.: Molecular interaction map of the mammalian cell cycle control and DNA repair systems. Molecular Biology of the Cell 10, 2703–2734 (1999)
3. Hucka, M., et al.: The systems biology markup language (SBML): A medium for representation and exchange of biochemical network models. Bioinformatics 19, 524–531 (2003)
4. Thomas, R., Gathoye, A.M., Lambert, L.: A complex control circuit: regulation of immunity in temperate bacteriophages. European Journal of Biochemistry 71, 211–227 (1976)
5. Kaufman, M., Soulé, C., Thomas, R.: A new necessary condition on interaction graphs for multistationarity. Journal of Theoretical Biology 248, 675–685 (2007)
6. Soulé, C.: Mathematical approaches to differentiation and gene regulation. C. R. Biologies 329, 13–20 (2006)
7. Soulé, C.: Graphic requirements for multistationarity. ComplexUs 1, 123–133 (2003)
8. Snoussi, E.: Necessary conditions for multistationarity and stable periodicity. J. Biol. Syst. 6, 3–9 (1998)
9. Gouzé, J.L.: Positive and negative circuits in dynamical systems. J. Biol. Syst. 6, 11–15 (1998)
10. Thomas, R.: On the relation between the logical structure of systems and their ability to generate multiple steady states or sustained oscillations. Springer Ser. Synergetics 9, 180–193 (1981)
11. Chabrier-Rivier, N., Chiaverini, M., Danos, V., Fages, F., Schächter, V.: Modeling and querying biochemical interaction networks. Theoretical Computer Science 325, 25–44 (2004)
12. Levchenko, A., Bruck, J., Sternberg, P.W.: Scaffold proteins biphasically affect the levels of mitogen-activated protein kinase signaling and reduce its threshold properties. PNAS 97, 5818–5823 (2000)

13. Ventura, A.C., Sepulchre, J.A., Merajver, S.D.: A hidden feedback in signaling cascades is revealed. PLoS Computational Biology (to appear, 2008)
14. Markevich, N.I., Hoek, J.B., Kholodenko, B.N.: Signaling switches and bistability arising from multisite phosphorylation in protein kinase cascades. Journal of Cell Biology 164, 353–359 (2005)
15. Calzone, L., Fages, F., Soliman, S.: BIOCHAM: An environment for modeling biological systems and formalizing experimental knowledge. BioInformatics 22, 1805–1807 (2006)
16. Fages, F., Soliman, S., Chabrier-Rivier, N.: Modelling and querying interaction networks in the biochemical abstract machine BIOCHAM. Journal of Biological Physics and Chemistry 4, 64–73 (2004)
17. Gillespie, D.T.: Exact stochastic simulation of coupled chemical reactions. Journal of Physical Chemistry 81, 2340–2361 (1977)
18. Qiao, L., Nachbar, R.B., Kevrekidis, I.G., Shvartsman, S.Y.: Bistability and oscillations in the huang-ferrell model of mapk signaling. PLoS Computational Biology 3, 1819–1826 (2007)
19. Ciliberto, A., Novák, B., Tyson, J.J.: Steady states and oscillations in the p53/mdm2 network. Cell Cycle 4, 488–493 (2005)

Rule-Based Modelling, Symmetries, Refinements

Vincent Danos[1,3], Jérôme Feret[2], Walter Fontana[2],
Russell Harmer[3], and Jean Krivine[2]

[1] University of Edinburgh
[2] Harvard Medical School
[3] CNRS, Université Paris Diderot

Abstract. Rule-based modelling is particularly effective for handling the highly combinatorial aspects of cellular signalling. The dynamics is described in terms of interactions between partial complexes, and the ability to write rules with such partial complexes -*i.e.*, not to have to specify all the traits of the entitities partaking in a reaction but just those that matter- is the key to obtaining compact descriptions of what otherwise could be nearly infinite dimensional dynamical systems. This also makes these descriptions easier to read, write and modify.

In the course of modelling a particular signalling system it will often happen that more traits matter in a given interaction than previously thought, and one will need to strengthen the conditions under which that interaction may happen. This is a process that we call *rule refinement* and which we set out in this paper to study. Specifically we present a method to refine rule sets in a way that preserves the implied stochastic semantics.

This stochastic semantics is dictated by the number of different ways in which a given rule can be applied to a system (obeying the mass action principle). The refinement formula we obtain explains how to refine rules and which choice of refined rates will lead to a neutral refinement, *i.e.*, one that has the same global activity as the original rule had (and therefore leaves the dynamics unchanged). It has a pleasing mathematical simplicity, and is reusable with little modification across many variants of stochastic graph rewriting. A particular case of the above is the derivation of a maximal refinement which is equivalent to a (possibly infinite) Petri net and can be useful to get a quick approximation of the dynamics and to calibrate models. As we show with examples, refinement is also useful to understand how different subpopulations contribute to the activity of a rule, and to modulate differentially their impact on that activity.

1 Semi-liquid Computing

To the eye of the computational scientist, cellular signalling looks like an intriguing computational medium. Various types of agents (proteins) of limited means interact in what, at first sight, may seem to be a liquid universe of chance encounters whre there is little causality. But in fact a rich decentralized choreography of bindings (complex formation) and mutual modifications (post-translational modifications) can be observed. Transient devices (complexes) are built by agents to

J. Fisher (Ed.): FMSB 2008, LNBI 5054, pp. 103–122, 2008.

integrate, convey, and amplify signals and channel them to the appropriate outputs (transcriptional regulation). The intricate pathways of the response to the
epidermal growth factor (EGF) sketched in Fig. 1 are a well-studied and well-
modelled example [1]. This universe of semi-liquid computing is brought about
by a surprisingly small number of elementary interactions. It sits somewhere in
between the worlds of the random graphs of statistical physics [2] which perhaps
lack expressivity, and the solid colliding sphere models of chemical kinetics [3]
which perhaps lack programmability.

The generativity of those systems, that is to say the number of different non-
isomorphic combinations (aka complexes or species) that may come to exist
along different realizations of the implied stochastic process, may well be enormous, but this does not say how complex those systems really are. A lot fewer
rules than there are reactions (interactions between complete complexes) may
be good enough to describe some of them. For instance the sketch of Fig. 1 once
properly formalized uses about 300 rules whereas it produces about 10^{40} unique
combinations. One sees that the number of rules is a more meaningful estimate
of its inherent complexity.

Rule-based languages [4,5,6,7,8,9,10,11], and more generally process algebraic
approaches to modelling [12,13,14,15,16,17,18,19], because they can express such
generic interactions, can work around this apparent descriptive complexity and
achieve compact descriptions. Let us also mention, although we will not treat
this aspect of the question here, that another benefit of rule-based modelling is
that one can trace the evolution of a system at the level of agents (or individuals)
and explore the causal relationships between events occurring in a system [6].

The difference between an assembly of agents with random uncorrelated encounters and a signalling system is that there is a causal structure channelling the
interactions towards a particular response. Typically a binding will not happen
before one or both of the bindees has been modified. Combining those micro-
causal constraints into a coherent pathway is a programming art that we don't
master or even understand yet, but one that signalling systems have been honing
for a considerable time. Rule-based modelling incorporates such causality constraints in the rules themselves by using partial complexes: not everything needs
to be described in a rule, only the aspects of the state of a complex which matter
for an event to happen need to be specified. That is the difference between a
reaction between complete entities, and a rule between partial ones. As said, this
reliance on partial complexes allows to capture compact descriptions and work
around the huge numbers of combinations one would have to contemplate (or
neglect) otherwise. The more detailed the partial complex, that is to say the less
partial, the more conditions must be met for a particular event to happen.

The purpose of the present paper is to understand better the mechanics of
refinement, that is to say the process by which one can make a complex less
partial, or equivalently a rule more demanding. We specifically consider the
problem of replacing a rule with a family of refined rules which will exhibit
the same collective activity, and will therefore generate an identical stochastic
behaviour. Note that there are really two questions in one: one is to define what

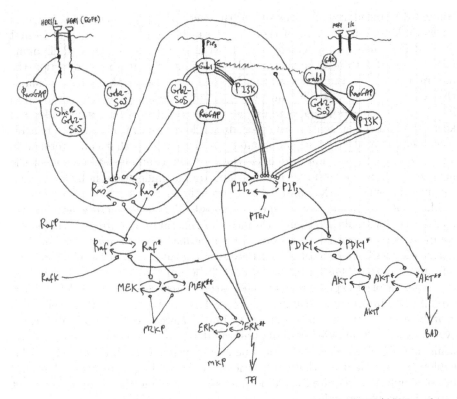

Fig. 1. A informal sketch of the many interactions involved in the ERK/AKT pathway responding to an EGF stimulus. The corresponding rule-based model generates about 10^{40} different species.

constitutes a good family of refined rules, another is to define their rates so as to preserve the underlying dynamics. It turns out that the latter question has an intimate connexion with the notion of symmetry, and what becomes of a symmetry group of a partial complex under refinement. The solution we propose to the former question can certainly be made to cover more cases (of which more later).

Seeing how the notion of partial complex is central to rule-based modelling[1] it certainly makes sense to try to theorize around it, as we start doing in this paper. But there are also very concrete reasons to do so. First it will often happen that in a modelling situation a rule has to be revised because people come to believe that its rate depends on more information about the context than the rule actually provides. A typical example would be that a post-translational modification increases or decreases the likelihood that an agent will bind another one. Replacing a rule with a bunch of more specific ones, in order to express those

---

[1] An aside: the name rule-based is a little unfortunate since just about any computational formalism is rule-based but that is the name under which this approach has become known in the biological modelling community.

context-dependent modulations of the rule activity, is a transformation which we call a kinetic refinement of the rule. It can be usefully decomposed as first introducing a neutral refinement -as defined in this paper- and second changing those base rates to achieve the modulation of interest. In this application the neutral refinement, that is to say the choice of refined rates that will *not* change the behaviour, serves as a baseline. One needs it to know where to start from. In fact even when one does not actually modulate the rates of the refined rules and keeps the refinement neutral, the procedure allows one to peek into the relative contributions of the various subpopulations of complexes that can intervene in an instance of the original unrefined rule (see the examples at the end of the next section). So for both reasons it is important to understand how to compute this baseline which is the question we address here.

Maximal refinements are of special interest. This is the case where one replaces a rule with all its ground instances (in general an infinite set) where only complete complexes take part. Such a transformation when applied to all rules in a rule set will obtain a set of multiset rewriting rules, that is to say a (possibly infinite) Petri net. This transformation will be unfeasible in general, owing to the combinatorial explosion mentioned earlier, because the obtained Petri net, even if finite in principle, will be simply too large to be written (this is not even a problem of computational complexity but of mere size of the output). However it is easy to imagine running truncated versions of a complete expansion using an ODE semantics. That could be useful for model calibration, and similar exploration mechanisms that are particularly demanding in terms of the number of simulations required while not necessarily needing the accuracy an exact expansion would provide.

We start with a brief presentation of Kappa which is the rule-based language we shall use in this paper. This is an occasion to get familiar with some of the notations, but is in no way a formal presentation. Then we turn to two simple examples of refinement to get a more concrete sense of what the notion of refinement is trying to achieve and how it is relevant to practical modelling questions. Explanations given in this paper, further than the ones given above, about the relevance of Kappa for the actual practical modelling are all to be found in the next Section. The reader interested in more can consult Refs. [6] and [7].

After this presentation the mathematical development reintroduces a simplified Kappa, this time in a completely formal and algebraic way which is conducive to a study of refinement which will be of general import and not tied in specific syntactical details. In fact the refinement formula we obtain is of general validity and assumes nothing about the arity of the rule to be refined, and actually assumes little about the rewriting framework itself. We are conscious that this incurs some cost to the reader unacquainted with basic category-theoretic notions in that some heavy-looking machinery is involved. However, mathematically it is natural, and hopefully at this stage the preliminary informal explanations will have clarified what is achieved in the mathematical development. After the derivation of the refinement formula for partial complexes first, and then for rules, we explain how to sharpen this result by considering more inclusive notions

of refinements. Only the first refinement formula is treated in detail; its extensions are just sketched.

2   A Brief Guide to Kappa

Let us start with an informal and brief account of our modelling language. In
Kappa, agents (think of them has idealized proteins) have sites and their sites
can be used to bind other sites (at most once), and can also hold an internal state.
The former possibility accounts for domain-mediated complex formation, while
the latter accounts for post-translational modifications. Accordingly one distin-
guishes three types of (atomic) rules for binding, unbinding, and modification. In
the full language one also considers agent creation and deletion (see later the formal presentation), and it is possible to combine actions in a single rule.

Note that a binding rule requires two distinct agents, each with a free (*i.e.*,
not already bound) site, which bind via those sites. In other words, it is not
possible to bind a site more than once.

A Kappa *model* consists of *(i)* an *initial solution* that declares the names and
all sites (with default state values for all sites we wish to carry a state) of the
relevant agents; and *(ii)* a *rule set* specifying how the initial solution may evolve.
We will see an example very soon.

The behaviour of a model is stochastic. Given a global state of the system
one assigns to each rule a likelihood to be applied which is proportional to the
number of ways in which this rule can be applied, and its intrinsic rate (the rate
is a measure of how efficient a rule is at turning a chance encounter of reagents
into an actual reaction). In the particular case where agents have no sites at all,
one has a Petri net, and the dynamics is none other than the mass action law
put in Gillespie form [3].

2.1   A Simple Cascade

As a way of getting more familiar with the notation we can consider a simple and
yet ubiquitous motif of cellular biology that consists of one protein (typically an
enzyme or kinase) covalently modifying another. Let us call them S (as signal)
and X which we are going to assume have each a unique site s. That situation
can be easily expressed by the following rule triplet:

$$S(s), X(s_u) \rightarrow S(s^1), X(s_u^1)$$
$$S(s^1), X(s^1) \rightarrow S(s), X(s)$$
$$S(s^1), X(s_u^1) \rightarrow S(s^1), X(s_p^1)$$

where we represent a binding between two sites by a shared exponent, s^1, and
the internal state of a site as a subscript to this site, s_u or s_p, where (u) p is a
mnemonic for (un-) phosphorylated.

One can add a second and similar triplet for X and a new agent Y:

$$X(s_p), Y(s_u) \rightarrow X(s_p^1), Y(s_u^1)$$
$$X(s^1), Y(s^1) \rightarrow X(s), Y(s)$$
$$X(s^1), Y(s_u^1) \rightarrow X(s^1), Y(s_p^1)$$

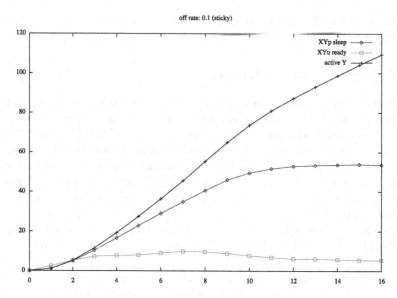

Fig. 2. Low off rate ($k = 0.1$): the activated Ys tend to stay attached to their activators X (the XYp 'sleep' curve dominates the XYu 'ready' one); as a consequence the production of activated Y is slowed down

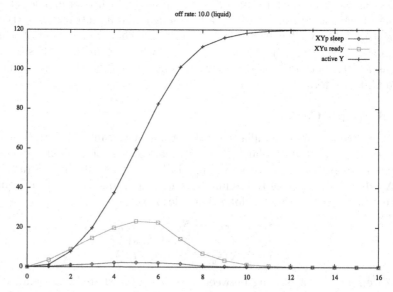

Fig. 3. Medium off rate ($k = 10$): most of the XY complexes have now an inactivated Y (the XYu 'ready' curve now dominates); the production of activated Y is visibly faster

Note that X has to be activated and Y inactivated for the binding and subsequent activation of Y to happen. This ensures in particular that no Y is activated in the absence of a signal S (an example of the causal constraints we were alluding to earlier). Such *cascades* regularly arise in real signalling networks. Referring back to the actual EGFR pathway in Fig. 1 we see the famous examples of the Ras, Raf, MEK and ERK cascade, and the PIP3, PDK1, and AKT one.

What about the dynamical behaviour of such a simplified cascade? We would like to understand how the cascade *throughput*, that is to say the rate of production of the active form of Y, depends on the rate k at which X detaches from Y (hereafter the XY *off-rate*), namely the rate of the rule $r := X(s^1), Y(s^1) \rightarrow X(s), Y(s)$ defined above.

Well intuitively, with too small an off-rate (high affinity binding), X will tend to remain bound to Y even after the Y has been activated. Whereas with too large an off-rate (low affinity binding), X will often detach from Y before having activated Y. Somewhere in between, an optimal choice of k will strike the right balance and maximise the rate of activation of Y.

This is something that we can verify numerically. Suppose one starts with $15S(s) + 60X(s_u) + 120Y(s_u)$ as an initial state, and suppose further all other rules have a rate of 1. As expected, we see in Fig. 2 where $k = 0.1$, that the activation of Y is slower than in Fig. 2, where $k = 10$; with an even higher $k = 1000$ the activation rate goes down again (Fig. 4).

This demonstrates the tension between binding loosely, and "not always getting the job done", and binding tightly which amounts to "sleeping on the job". It also nicely shows that the cascade throughput depends on a lot more than just the rate attached to the rule performing that activation.

Now our off-rate k measures how likely it is that X and Y will detach - *independently* of their respective internal states. If we were to optimize the cascade throughput it would be natural to let the off-rate depend on the state of Y. In terms of rules, all we have to do is to split the unbinding rule under consideration into two subcases:

$$r_u := X(s^1), Y(s_u^1) \rightarrow_{k_u} X(s), Y(s_u)$$
$$r_p := X(s^1), Y(s_p^1) \rightarrow_{k_p} X(s), Y(s_p)$$

with respective rates k_u, k_p. One calls the substitution of r with such more specific rules r_u, r_p a *refinement*. If in addition the new rates are both taken equal to k, then in this simple case evidently the behaviour of the system will be unchanged. That special case, where nothing changes in the dynamics, is what we call a neutral refinement.

If one favours unbinding from active Y, then clearly this allows X both to bind long enough to Y to activate it and then to unbind quickly to maximize throughput. In particular the combination $k_p = \infty$ (detach as soon as activated), and $k_u = 0$ (never detach before activation) leads to the best possible throughput, all other things being equal (see Fig. 5).

2.2   A Less Obvious Refinement

Here is a second example which shows that choosing the rates of the refined rules and obtaining a neutral refinement may require some ingenuity.

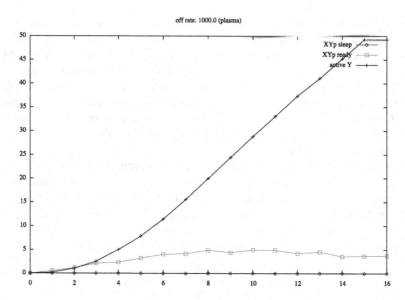

Fig. 4. High off rate ($k = 1000$): the production of activated Y has gone down again

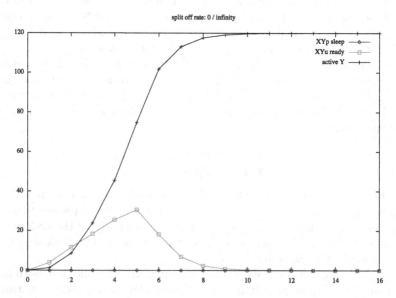

Fig. 5. Split rate ($k_u = 0$, $k_p = \infty$): there is no XYp anymore; the production of activated Y is optimal

Consider two agent types B, C each with only one site x, and define a family of systems $x(n_1, n_2)$ consisting of n_1 single Cs and n_2 dimers $C(x^1), B(x^1)$. In other words set $x(n_1, n_2) := n_1C(x) + n_2(C(x^1), B(x^1))$. Now consider the rule $r := C(), B() \rightarrow_1 C()$ with rate 1. Note that r does not mention x at all (we say that both agents have an empty signature in this case). This means that r applies irrespective of the binding state of x in B and C. Both agents could be free, or bound, or even bound together. Whichever is the case, the effect of the rule will be the same, namely to delete a B and to bring $x(n_1, n_2)$ to a new state $x(n_1 + 1, n_2 - 1)$. This supposes $n_2 > 0$. If on the other hand $n_2 = 0$ then there is no B left in the system and no further event is possible (deadlock).

Now we would like to refine r into mutually exclusive sub-cases depending on the relationship in which C and B stand; specifically we want to use the following three refined rules:

$$r_1 := C(x^1), B(x^1) \rightarrow_1 C(x)$$
$$r_2 := C(x^1), B(x^1), C(x^2), B(x^2) \rightarrow_2 C(x), C(x^2), B(x^2)$$
$$r_3 := C(x^1), B(x^1), C(x) \rightarrow_1 C(x), C(x)$$

Each of them is a particular case of r in the sense that their left hand sides embed (sometimes in more than one way) that of r (see below the notion of morphism). Intuitively, r_1 is the sub-case where B, C are bound together, r_2 is the sub-case where they are both bound but not together, and r_3 is the sub-case where B is bound but C is free. Given the particular family of states $x(n_1, n_2)$ we are dealing with, those seem to cover all possible cases, and to be indeed exclusive.

Define the *activity* of a rule as the number of possible ways to apply the rule multiplied by its rate. This determines its likelihood to apply next and only depends on the current state of the system. Now we have chosen for each refined rule a rate (indicated as a subscript to the reaction arrow), and in particular r_2 was assigned a rate of 2. We claim this is the unique correct choice if one wants the stochastic behaviour of the system to be preserved by the refinement. Figure 6 shows a run of the refined system with $x(0, 100)$ as the initial state. The y axis traces the activity of all rules including the base one r.

We see that indeed at all times the refined activities add up to the original one (the top curve).

There are other things worth noticing. Firstly, r_1 keeps a low probability that decreases linearly during the simulation since its activity is exactly the number of dimers n_2; so suppressing r_1 would change very little to the behaviour of the system. Secondly, r_2 dominates the early events, since near the initial state there are only dimers, and no free Cs yet; however, as time passes there will be more of those free Cs, and the corresponding rule r_3 will come to dominate. Hence we see that the relative importance of the sub-cases changes over time, and that refinement can be used as a way of profiling the contribution various subpopulations of agents make to a given type of event.

The corrective factor applied to r_2 accounts for two opposite effects: on the one hand r_2 embeds r in more than one way which tends to scale the rate of r_2 upwards, on the other hand r_2 is more symmetric than r and that would tend to scale the rate of r_2 downwards.

What we are interested in is to handle the general case, *i.e.*, to explain what constitutes a good set of refined rules as r_1, r_2, and r_3 above, and how one can

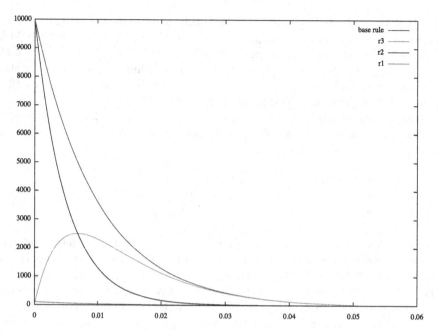

Fig. 6. The activities of the refined rules r_1, r_2, and r_3 add up exactly to that of the initial rule r (top curve)

choose the refined rates in a way that the global activity is preserved. We will return to the example once we have a general solution.

3   Rule-Based Modelling

To give proper generality and portability to our study, we will frame it into some simple categorical language where a system is seen as an object x and the various ways a rule r may apply to x are identified using a notion of morphism f from r's left hand side to x.

As said we shall also simplify the Kappa syntax in two respects. First, we suppose agents have no internal states. Second, we suppose no wildcards are used in left hand sides, e.g., expressions like $A(x^-)$ meaning x is bound to some unspecified other site, are not considered. The former simplification is only a matter of readability, as internal values offer no difficulty. The latter simplification is more significant, and we will see later in our development that reintroducing wildcards allows us to strengthen our main result. With these simplifications we can give a syntax-less presentation of Kappa that will facilitate the derivation of the refinement formula.

We suppose given two sets $\mathcal{A}$ and $\mathcal{S}$ of agent names and sites.

A *matching* over a set X is an irreflexive and symmetric binary relation on X such that no element is in relation with more than one element.

Definition 1. *An object is a quadruple* $(V, \lambda, \sigma, \mu)$ *where:*
- *V is a set of nodes,*
- *$\lambda \in \mathcal{A}^V$ assigns names to nodes,*
- *$\sigma \in \mathcal{P}(\mathcal{S})^V$ assigns sets of sites to nodes,*
- *μ is a matching over $\sum_{v \in V} \sigma(v)$.*

The matching represents bindings, and hence any given site can be bound at most once. A node however can be bound many times via different sites.

We define $(u, x) \in \mu$ as shorthand for $\exists (v, y) : (u, x, y, v) \in \mu$, and say u, x is *free* when $(u, x) \notin \mu$, *bound* when $(u, x) \in \mu$.

The simplest non-empty object is a single node named A with no sites and therefore no binding. In the preceding section we wrote $A()$ for this object. There we also introduced a textual notation to designate objects where bindings are indicated by exponents.

Note that we sometimes use the same family of symbols x, y, etc. for sites and objects. Hopefully this will not cause any confusion since they are entities of a very different nature.

We define a *signature* as a map $\Sigma : \mathcal{A} \to \mathcal{S}$; this can be used to constrain the set of sites per agent type. We write $x \leq \Sigma$ if for all $v \in V_x$, $\sigma_x(v) \subseteq \Sigma(\lambda_x(v))$; likewise we write $\Sigma \leq x$ if for all $v \in V_x$, $\Sigma(\lambda_x(v)) \subseteq \sigma_x(v)$, and $x : \Sigma$ when $x \leq \Sigma \leq x$.

When $x : \Sigma$ for some Σ, we say x is *homogeneous*, which means all agents of the same type in x use exactly the same set of sites.

Definition 2. *An arrow* $(V, \lambda, \sigma, \mu) \to (V', \lambda', \sigma', \mu')$ *is a map* $f : V \to V'$ *such that*
- *1) f preserve names: $\lambda' \circ f = \lambda$*
- *2) f preserve sites: $\sigma' \circ f \supseteq \sigma$*
- *3a) f preserve edges: $(u, x, y, v) \in \mu \Rightarrow (f(u), x, y, f(v)) \in \mu'$*
- *3b) f reflects edges: $(f(u), x) \in \mu', x \in \sigma_x(u) \Rightarrow (u, x) \in \mu$*
- *4) f is a monomorphism*

This then is the category of *graphs with sites* we shall work with. We also call arrows morphisms sometimes; we write $[x, y]$ for the arrows from x to y; $iso[x, y]$ for the isomorphisms (meaning invertible arrows), and therefore $[x, x] = iso[x, x]$ denotes the set of *automorphisms* (or symmetries) of x; we say that y *embeds* x when $[x, y] \neq \varnothing$.

Define the *image* of $f \in [x, y]$ as $Im(f) := \{f(v), x; v \in V, x \in \sigma(v)\}$.

Note that $Im(f)$ is but a subset of $\sum_{v \in V} \sigma'(f(v))$, and only sites in $Im(f)$ are mentioned in the arrow-defining clauses above.

One has obviously a forgetful functor to the category of graphs and graph morphisms, and that allows us to import the usual graph-theoretical vocabulary of connected components and paths, which we will freely use in the sequel. Note that, from the point of view of graphs, the reflectivity condition 3b) above does not really make sense, one really needs sites to express edge reflection. Moreover the rather stringent notion of arrow constrains the homsets $[x, y]$:

Lemma 1 (rigidity). *Suppose x is connected, then any non-empty partial injection f from V_x to V_y extends to at most one morphism in $[x, y]$.*

Proof: If f is strictly partial, that is to say $V_x \setminus dom(f)$ is not empty, pick a v in there such that for some node $w \in dom(f)$, and some sites x, y, $(w, y, v, x) \in \mu_x$. This is always possible because x is connected. Then, either $(f(w), y, v', x) \in \mu_y$ for some $v' \in V_y$, and by 3a) one must extend f as $f(v) = v'$, or there is no such extension.   □

Clearly being a monomorphism, *i.e.*, being post-cancellable, is equivalent to being a one-one map. On the other hand there are far more epimorphisms than surjections:

Lemma 2 (epis). *A map $h \in [x, y]$ is an epimorphism iff every connected component of y intersects $f(x)$; that is to say for all connected component $c_y \subseteq y$, $h^{-1}(c_y) \neq \varnothing$.*

Proof: Suppose $f_1 h = f_2 h$ for $h \in [x, y]$, $f_i \in [y, z]$, and let $c_y \subseteq y$ be a connected component of y such that $h^{-1}(c_y) \neq \varnothing$. Pick u such that $h(u) \in c_y$, then $f_1(h(u)) = f_2(h(u))$ and by the preceding lemma $f_1/c_y = f_2/c_y$.   □

We write $[x, y]^e \subseteq [x, y]$ for the epis from x to y.

4   Object Refinements

Now that we have our basics in place we turn to the first question of what constitutes a refinement of a (partial) object s. As we have seen in the example, a refinement of s is intuitively a collection of objects t_i that embed s and such that any embedding in an object of higher signature x (*i.e.*, that has more sites everywhere) can be unambiguously attributed to one t_i. We first make this intuition into a real definition and then proceed to define the refinements of rules.

Definition 3 (factorisation). *One says an object t factors $f \in [s, x]$ iff $f = \gamma\phi$ for some $\phi, \gamma \in [s, t]^e \times [t, x]$; ϕ, γ is called a factorisation of f via t.*

The first thing to notice is that one cannot ask for unique factorisations.

   Suppose given a factorisation $\phi, \gamma \in [s, t]^e \times [t, x]$ of f via t and an *isomorphism* $\alpha \in [t, t']$. Define $\phi', \gamma' := \alpha\phi, \gamma\alpha^{-1} \in [s, t'] \times [t', x]$; this new pair verifies $\gamma\phi = \gamma'\phi'$, and since ϕ' is clearly an epimorphism, the pair is also a factorisation of f via t'.

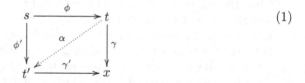

$$(1)$$

In this case we will say that ϕ, γ and ϕ', γ' are *conjugate* under α, and write $\phi, \gamma \simeq_{tt'} \phi', \gamma'$. We also write $[s, t] \times_{[t, t]} [t, x]$ for the quotient of $[s, t] \times [t, x]$ under $\simeq_{tt}$; this notation is justified by the following:

Lemma 3 (conjugates). *The equivalence relation $\simeq_{tt}$ has $|[s,t] \times [t,x]|/|[t,t]|$ classes.*

Proof: Suppose, using the notations of (1), that $\phi, \gamma \simeq_{tt'} \phi', \gamma'$, then this uniquely determines α since $\gamma\alpha^{-1} = \gamma\alpha'^{-1}$ implies $\alpha = \alpha'$ by γ being a monomorphism. In particular the set of conjugates of ϕ, γ over the same t is in one-one correspondence with $[t,t]$. $\qquad\square$

Note that being an epimorphism is stable by conjugation so we can say that a class is an epimorphism, and we can restrict the equivalence to $[s,t]^e \times [t,x]$, so the version of the Lemma relative to $[s,t]^e$ also holds.[2]

Unicity of factorization is then to be understood up to isomorphisms; furthermore, even if we select one representative t_i per isomorphism class, unicity is up to automorphisms of each of the representative t_i.

Definition 4 (object refinement). *Given s, Σ such that $s \leq \Sigma$, a refinement of s under Σ, written $\Sigma(s)$, is a set of objects obtained by selecting one representative in each isomorphism class defined by $\{t \mid t : \Sigma, [s,t]^e \neq \varnothing\}$.*

Note that the actual choice of representatives does not matter, but we do have to choose one for our counting purposes.

Another noteworthy fact is that $\Sigma(s)$ in general will be infinite. However in practice one may get information about the reachables of the system which will allow to control the size of the expansion [4]; indeed it is not necessary to include ts which are not reachable, and we took advantage of this in the example of the first section.

Lemma 4 (injectivity). *Given Σ, s, x such that $s \leq \Sigma \leq x$ the composition map from the disjoint sum $\sum_{t\in\Sigma(s)} [s,t]^e \times_{[t,t]} [t,x]$ to $[s,x]$ is injective.*

Proof: Suppose given two factorisations $f = \gamma\phi = \gamma'\phi'$ via t and t' as in (1).

$$v \in s \xrightarrow{\quad\phi\quad} t \supseteq c \ni \phi(v) \qquad\qquad (2)$$

$$\phi'(v) \in c' \subseteq t' \xrightarrow{\quad\gamma'\quad} x$$

Pick a connected component $c \subseteq t$, such that $\phi(v) \in c$ for some $v \in s$. Call $c' \subseteq t'$ the connected component of $\phi'(v)$ in t'. By construction $\gamma(c)$ and $\gamma'(c')$ intersect

---

[2] This prompts a more general argument for the restriction of that Lemma to the case of $[s,t]^e$. Suppose given ϕ, γ as above, one has a map from $[t,t]$ to the class $[\phi,\gamma]$ in $[s,t] \times [t,x]$: $\alpha \mapsto \alpha \cdot (\phi,\gamma) := \phi\alpha, \alpha^{-1}\gamma$ (an action of the group $[t,t]$ on $[s,t] \times [t,x]$); this map is surjective by definition of conjugation; it is also injective because γ is a mono. Now if in addition ϕ is an epi, it is injective for a second reason, namely ϕ is an epi ($\alpha_1\phi = \alpha_2\phi$ implies $\alpha_1 = \alpha_2$). This seems to indicate that one can relax the monomorphism requirement in the ambient category and still develop the same theory.

at $\gamma\phi(v) = f(v) = \gamma'\phi'(v)$. It is easy to see that they both are Σ-homogeneous. This means they must be equal.

Indeed suppose $w \in \gamma(c)$ is a node which is directly connected to $\gamma(c) \cap \gamma'(c')$, meaning w is such that $(u, x, y, w) \in \mu_x$, for some $u \in \gamma(c) \cap \gamma'(c')$ and $(u, x), (w, y) \in Im(\gamma)$. Because c' is Σ-homogeneous, $u, x \in Im(\gamma')$, ie x is also a site of the (unique) antecedent of u in c', which we can write $x \in \sigma_{t'}\gamma'^{-1}(u)$. By condition 3b) this site cannot be free, and by 3a) it must be bound to $\gamma'^{-1}(w), y$, so $w \in \gamma'(c')$. Since $\gamma(c)$ is connected, $\gamma(c) \cap \gamma'(c')$ must contain $\gamma(c)$, and by symmetry $\gamma'(c')$.

Hence $\gamma(c) = \gamma'(c')$, therefore c and c' are isomorphic. In fact, since ϕ is an epi, we can repeat the above for any connected component in t, and therefore t embeds in t' (it is readily seen that the assignment of a c' to a c above is injective), and by symmetry they must be isomorphic under a certain isomorphism α. By definition of $\Sigma(s)$ we have picked exactly one representative in each isomorphism class, therefore $t = t'$, $\alpha \in [t, t]$, and the two factorizations are conjugate under α. $\square$

Theorem 1. *Given Σ, s, x such that $s \leq \Sigma$ and $x : \Sigma$:*

$$[s, x] \simeq \sum_{t \in \Sigma(s)} [s, t]^e \times_{[t,t]} [t, x]$$

Proof: From the preceding lemma we know the composition map is injective, so all there remains to prove is that it is surjective.

Consider $f \in [s, x]$, define $f(s) := \{u \mid \exists x : (u, x) \in Im(f)\} \subseteq V_x$, and write $[f(s)]$ for the connected closure of $f(s)$ in x. We claim there is a $t \in \Sigma(s)$ which is isomorphic to $[f(s)]$. Indeed every node in $[f(s)]$ has a signature in accordance with Σ because $x : \Sigma$, and $[f(s)]$ embeds s since $f(s)$ does (via f). $\square$

Using Lemma 3 in addition we can use the above theorem to obtain:

Corollary 1. *Given Σ, s, x such that $s \leq \Sigma$ and $x : \Sigma$, one has:*

$$||[s, x]|| = \sum_{t \in \Sigma(s)} |[s, t]^e|/|[t, t]| \cdot |[t, x]| \qquad (3)$$

There are several things worth noticing about the theorem and its numerical form as a corollary.

First, the $|[s, t]^e|/|[t, t]|$ is a *static* term that can be computed once and for all, and which we shall use to determine the rule rates. The positive contribution $[s, t]^e$ is rather intuitive since the more copies of s one finds in t the higher the contribution of t to the number of instances of s should be; the negative contribution $|[t, t]|$ is less intuitive however.

Second, one cannot relax the homogeneity condition on x and ask only $\Sigma \leq x$. That would break the easy part of the proof, namely that of surjectivity. Here is an example; set $s := A(x) < \Sigma := A \mapsto \{x, y\} < A(x, y^1, z), A(x, y, z^1) =: x$. Choose f to be the 'left' morphism mapping s's unique A to $A(x, y^1, z)$ in x; then $[f(s)] = x$ and no $t \in \Sigma(s)$ can factorize f because the (y, z) binding is not reproducible in t, because $z \notin \Sigma(A)$.

However, one can modify the notion of object (and accordingly that of arrow) by introducing new partial objects such as $t = A(x, y^{\neg \Sigma})$, meaning y binds an otherwise unspecified non-Σ site (ie A, y is bound to some B, z such that $z \notin \Sigma(B)$). This t is homogeneous and factorizes the f above. This variant allows to recover surjectivity and extend our decomposition theorem above. Similar wildcard expressions are already present in the actual syntax of Kappa, and it is amusing to see that those convenient notations also have a theoretical status.

This begs a last remark, namely that we are the ones choosing how to relate the base object s and its refinement. For example, here, we are using epis to relate them. Below we will allude to a finer-grained correspondence based on using a pointed version of the ambient category that will allow us to go beyond the homogeneity requirement in another way. But before we do that we will return to the example of the first section.

4.1   Example Continued

We can now reconsider our initial example. Set $s := C(), B()$, for the left hand side of the base rule r, and t_i for that of the refined rule r_i:

$$t_1 := C(x^1), B(x^1)$$
$$t_2 := C(x^1), B(x^1), C(x^2), B(x^2)$$
$$t_3 := C(x^1), B(x^1), C(x)$$

Set also $\Sigma := B, C \mapsto \{x\}$. Clearly $s < \Sigma$ and t_i, and $x(n_1, n_2)$ are Σ-homogeneous. Besides due to the particular form of $x(n_1, n_2)$, the t_is are the only elements in $\Sigma(s)$ that $x(n_1, n_2)$ embed. Using Lemma 3 we get:

$$|[s, x(n_1, n_2)]| = n_2(n_1 + n_2)$$
$$|[s, t_1]^e \times_{[t_1,t_1]} [t_1, x(n_1, n_2)]| = |[s, t_1]^e [t_1, x(n_1, n_2)]|/|[t_1, t_1]| = 1.n_2/1 = n_2$$
$$|[s, t_2]^e \times_{[t_2,t_1]} [t_2, x(n_1, n_2)]| = |[s, t_2]^e [t_2, x(n_1, n_2)]|/|[t_2, t_2]| = 2.n_2(n_2 - 1)/2$$
$$|[s, t_3]^e \times_{[t_3,t_1]} [t_3, x(n_1, n_2)]| = |[s, t_3]^e [t_3, x(n_1, n_2)]|/|[t_2, t_2]| = 1.n_1 n_2/1 = n_1 n_2$$

and the corollary correctly predicts $n_1 n_2 + n_2(n_2 - 1) + n_2 = n_1(n_1 + n_2)$.

4.2   Pointed Refinements

Let us look at an example which breaks injectivity. This is the kind of complication the theorem is staying cautiously away from by asking the ts to be homogeneous.

The set of nodes $V_s = \{1, 2\}$ is represented as subscripts to agents below; the subscripts to the y sites, y_0 and y_1, denote bindings to agents with only one site and different names (to save space):

$$s = A(x^1)_1, A(x^1)_2 \xrightarrow{\quad I \quad} t_0 = A(x^1, y_0)_1, A(x^1)_2$$
$$\downarrow I \qquad\qquad\qquad I \qquad\qquad \downarrow I$$
$$t_1 = A(x^1)_1, A(x^1, y_1)_2 \xrightarrow{\quad I \quad} x = A(x^1, y_0)_1, A(x^1, y_1)_2$$

If one refers to the situation of (1), the unique possible candidate conjugating α, i.e., the unique diagonal that makes both triangle commute, fails to be a morphism. That means that t_0, t_1 provide really distinct extensions of $f(s)$ in x and form an ambiguous decomposition of s. Indeed, applying (wrongly since the t_is are not homogeneous) the refinement formula (3) betrays this redundancy problem since $|[s, x]| = 2$ while $|[s, t_i]|/|[t_i, t_i]||[t_i, x]| = 2$.

To deal with a case such as this one, one needs to break the symmetry. To do this, a possibility is to work out the static part of the refinement formula in a *pointed* subcategory where objects have in addition to their usual structure a distinguished node per connected component, and arrows are asked to preserve them. Then one can replace homogeneity by a weaker requirement, namely that across all expansions of s no two agents with the same coordinates with respect to a distinguished node differ in their signature. In the example above, that would force to decide whether the additional binding is to sit on the distinguished node or not, and *then* both extensions would become truely distinct and unambiguous. Obviously a little more work is needed to say with complete confidence that this will work, but it seems it will.

5   Rule Refinements

Now that we know how to refine objects, we will proceed to the case of rules.

5.1   Action, Rules, Events

An atomic action on s is one of the following:
- an edge addition $+(u, x, y, v)$
- an edge deletion $-(u, x, y, v)$
- an agent addition $+(A, \sigma)$ with A a name, σ a set of free sites
- an agent deletion $-(u)$ with $u \in V_s$, $v \in V_s$, $x \in \sigma_s(u)$, and $y \in \sigma_s(v)$.

An action on s is a finite sequence of atomic actions on s. An atomic action is well defined on s:
- if $\alpha = +(u, x, y, v)$, when both (u, x) and (v, y) are free in s,
- if $\alpha = -(u, x, y, v)$, when $(u, x, y, v) \in \mu_s$.

This notion extends readily to non-atomic actions; we consider only well-defined actions hereafter.

Definition 5. *A rule is a triple $r = s, \alpha, \tau$ where:*
- *s in an object,*
- *α is an action on s,*
- *and τ a rate which can be any positive real number.*

We write $\alpha \cdot s$ for the effect of the action α on s.

Given $f \in [s, x]$ and α there is an obvious definition of the transport of α along f, written $f(\alpha)$, and it is easy to verify that $f(\alpha)$ is itself a well-defined action on x if α is a well-defined action on s (condition 3b) is crucial though).

Definition 6. *A set R of rules defines a labelled transition relation:*

$$x \longrightarrow_f^{s,\alpha,\tau} f(\alpha) \cdot x \qquad (4)$$

where $s, \alpha, \tau \in R$, and $f \in [s, x]$.

The labelled transition system just defined can be enriched quantitatively in a way that generalizes the notion of stochastic Petri nets [20] (Petri nets correspond to the case of a uniformly empty signature $\Sigma = \varnothing$).

To do this we need to define the activity of a rule.

Definition 7. *Given an object x and a rule $r = s, \alpha, \tau$, the activity of r at x is $a(x, r) := \tau \lvert [s, x] \rvert$, and the global activity of a set of rules R at x is $a(x) := \sum_{r \in R} a(x, r)$.*

Supposing $a(x) > 0$, the probability at x that the next event is $f \in [s, x]$ is $p(x, f) := \tau/a(x)$, and the subsequent time advance is a random variable $\delta t(x)$ such that $p(\delta t(x) > t) := e^{-a(x)t}$. For our present purposes, all we need to remember is that the quantitative structure of the transition system is entirely determined by the activities of its rules. In fact this means our result will hold for a larger class of stochastic system -for what it is worth.

5.2   The Main Result

Given a rule $r = s, \alpha, \tau$ and $\theta \in [s, t]$, we define $\theta(r) := \theta(s), \theta(\alpha), \tau$.

We say r, r' are isomorphic rules, written $r \simeq r'$, if there is an isomorphism $\theta \in [s, s']$ such that $r' = \theta(r)$. If that is the case then r and $\theta(r)$ have isomorphic transitions:

$$x \longrightarrow_{f \in [s,x]}^{r} f(\alpha) \cdot x \Leftrightarrow x \longrightarrow_{f \theta^{-1} \in [\theta(s),x]}^{\theta(r)} f \theta^{-1}(\theta(\alpha)) \cdot x$$

and in particular the same activity $a(r, x) = a(\theta(r), x)$.

Definition 8 (rule refinement). *Given s, Σ such that $s \leq \Sigma$ and $r = s, \alpha, \tau$, the refinement of r under Σ is the following family of rules:*

$$\Sigma(s, \alpha, \tau) := (t, \phi(\alpha), \tau; t \in \Sigma(s), \phi \in [s, t]^e / [t, t]) \qquad (5)$$

where the notation $\phi \in [s, t]^e / [t, t]$ means that for each t, one selects one $\phi \in [s, t]^e$ per symmetry class on t (the equivalence relation $\exists \theta \in [t, t] : \phi = \theta \phi'$).

It is easily seen that the particular selection made is irrelevant, but one has to choose one to define refinement as a syntactic transformation.

Note also that the above family can have isomorphic or even identical rules, it is important to have them all, *i.e.*, *stricto sensu* the expansion is a multiset of rules not a set. However one can always pack n isomorphic copies together by choosing a representative and multiplying its rate by n so we carry on with our slight abuse of terminology.

Given R a rule set, r a rule in R, we write $R[r \backslash \Sigma(r)]$ for the rule set obtained by replacing r with $\Sigma(r)$.

We write $r = s, \alpha, \tau \leq \Sigma$ if $s \leq \Sigma$, and $R \leq \Sigma$ if for all $r \in R$, $r \leq \Sigma$.

Theorem 2. *Given R, Σ, such that $R \leq \Sigma$, one has $R[r \backslash \Sigma(r)] \leq \Sigma$, and R and $R[r \backslash \Sigma(r)]$ determine the same stochastic transition system over Σ-homogeneous objects.*

Proof: By Th. 1 events $f \in [s,x]$ associated to rule $r = s, \alpha, \tau$ are in one-one correspondence with factorizations $f = \gamma \phi$ via some t, and therefore determine a unique matching refined event γ. This refined event has the same effect as f since:

$$x \xrightarrow[\gamma \in [t = \phi(s), x]]{t, \phi, \tau} \gamma \phi(\alpha) \cdot x = f(\alpha) \cdot x$$

so r and its refinements are equally likely and have the same effect on the underlying state x; hence their stochastic transition systems are the same. $\qquad\square$

Note that the activity of t, ϕ, τ in the refined system is $\tau |[t, x]|$ so the cumulative activity of the refined rules is:

$$\sum_{t \in \Sigma(s)} \sum_{\phi \in [s,t]^e/[t,t]} \tau |[t,x]| = \sum_{t \in \Sigma(s)} \tau |[s,t]^e|/|[t,t]||[t,x]| = a(r,x)$$

by Coro. 1, so we can directly derive the fact that the refined rules have the same activity, but we also needed to prove they have the same effect.

5.3   Example Concluded

We can now conclude our initial example.

There we had $s := C(), B()$, and:

$$t_1 := C(x^1), B(x^1)$$
$$t_2 := C(x^1), B(x^1), C(x^2), B(x^2)$$
$$t_3 := C(x^1), B(x^1), C(x)$$

Since $|[s, t_2]^e| = 2$ (recall that epis must have images in all connected components), the refinement of r via t_2 will contribute two rules to $\Sigma(r)$ -according to Def. 8. In this particular case the action of the rule to be refined is $\alpha(r) = -B$, and both epimorphisms $\phi \in [s, t_2]$ lead to the same transported action $\phi(-B)$ up to isomorphism. One can then pack them into one rule r_2, as we did intuitively when we considered the example, and as a consequence the rates must be added. This explains why r_2 has a rate of 2.

6   Conclusion

We have presented in this article the beginning of a theory of refinements for rule-based modelling. Specifically we have defined what constitutes a notion of a good set of refined rules and how, given such a set, one can compute the new refined rates in such a way that the overall activity of the system is preserved and the underlying stochastic semantics therefore unchanged. We have suggested two improvements to extend the type of refinement one can consider.

We have also shown how one can use such refinements to obtain a complete expansion (at least in principle), a construction which could be useful in practice to get cheap and fast approximations of a system. We have further shown by examples that refinements can be useful to modulate the influence of the context in which a rule is applied.

A point worth commenting in this conclusion is that the formulas obtained in our two main results, Th. 1 and 2, are couched in rather general terms and are likely to be of a larger relevance than the particular case of graph-rewriting we were contemplating here. In particular the epi-mono factorization system which we rely on implicitly for the concrete case we have treated would point to a more abstract approach. That in itself is valuable since such combinatorial results as we have presented here can become nearly intractable if looked at in a too concrete way. This in fact is one of the reasons why we framed our results in a categorical language which has revealed the pervasiveness of symmetries (the other reason is that the syntax is simpler to deal with). It would be particularly interesting to recast the theory in the axiomatic framework of adhesive categories [21], with a view on understanding the formula as a traditional partition formula (which it is, at least intuitively).

A longer term goal that this preliminary work might help to reach is that of finding exact model reduction techniques. This needs to lift a key assumption made here, namely that refinements are made of mutually exclusive sub-cases.

References

1. Orton, R.J., Sturm, O.E., Vyshemirsky, V., Calder, M., Gilbert, D.R., Kolch, W.: Computational modelling of the receptor tyrosine kinase activated MAPK pathway. Biochemical Journal 392(2), 249–261 (2005)
2. Söderberg, B.: General formalism for inhomogeneous random graphs. Physical Review E 66(6), 66121 (2002)
3. Gillespie, D.T.: Exact stochastic simulation of coupled chemical reactions. J. Phys. Chem. 81, 2340–2361 (1977)
4. Danos, V., Feret, J., Fontana, W., Krivine, J.: Abstract interpretation of cellular signalling networks. In: Logozzo, F., et al. (eds.) VMCAI 2008. LNCS, vol. 4905, pp. 83–97. Springer, Heidelberg (2008)
5. Danos, V., Feret, J., Fontana, W., Krivine, J.: Scalable simulation of cellular signaling networks. In: Shao, Z. (ed.) APLAS 2007. LNCS, vol. 4807, pp. 139–157. Springer, Heidelberg (2007)
6. Danos, V., Feret, J., Fontana, W., Harmer, R., Krivine, J.: Rule-based modelling of cellular signalling. In: Caires, L., Vasconcelos, V.T. (eds.) CONCUR. LNCS, vol. 4703. Springer, Heidelberg (2007)
7. Danos, V.: Agile modelling of cellular signalling. In: Proceedings of ICCMSE (2007)
8. Blinov, M.L., Faeder, J.R., Goldstein, B., Hlavacek, W.S.: A network model of early events in epidermal growth factor receptor signaling that accounts for combinatorial complexity. BioSystems 83, 136–151 (2006)
9. Hlavacek, W.S., Faeder, J.R., Blinov, M.L., Posner, R.G., Hucka, M., Fontana, W.: Rules for Modeling Signal-Transduction Systems. Science's STKE 2006(344) (2006)
10. Blinov, M.L., Yang, J., Faeder, J.R., Hlavacek, W.S.: Graph theory for rule-based modeling of biochemical networks. In: Abadi, M., de Alfaro, L. (eds.) CONCUR 2005. LNCS, vol. 3653. Springer, Heidelberg (2005)

11. Faeder, J.R., Blinov, M.L., Goldstein, B., Hlavacek, W.S.: Combinatorial complexity and dynamical restriction of network flows in signal transduction. Systems Biology 2(1), 5–15 (2005)
12. Regev, A., Silverman, W., Shapiro, E.: Representation and simulation of biochemical processes using the π-calculus process algebra. In: Altman, R.B., Dunker, A.K., Hunter, L., Klein, T.E. (eds.) Pacific Symposium on Biocomputing, vol. 6, pp. 459–470. World Scientific Press, Singapore (2001)
13. Priami, C., Regev, A., Shapiro, E., Silverman, W.: Application of a stochastic name-passing calculus to representation and simulation of molecular processes. Information Processing Letters (2001)
14. Regev, A., Shapiro, E.: Cells as computation. Nature 419 (September 2002)
15. Priami, C., Quaglia, P.: Beta binders for biological interactions. In: Danos, V., Schachter, V. (eds.) CMSB 2004. LNCS (LNBI), vol. 3082, pp. 20–33. Springer, Heidelberg (2005)
16. Danos, V., Krivine, J.: Formal molecular biology done in CCS. In: Proceedings of BIO-CONCUR 2003, Marseille, France. Electronic Notes in Theoretical Computer Science, vol. 180, pp. 31–49. Elsevier, Amsterdam (2003)
17. Regev, A., Panina, E.M., Silverman, W., Cardelli, L., Shapiro, E.: BioAmbients: an abstraction for biological compartments. Theoretical Computer Science 325(1), 141–167 (2004)
18. Cardelli, L.: Brane calculi. In: Proceedings of BIO-CONCUR 2003. Electronic Notes in Theoretical Computer Science, vol. 180. Elsevier, Amsterdam (2003)
19. Calder, M., Gilmore, S., Hillston, J.: Modelling the influence of RKIP on the ERK signalling pathway using the stochastic process algebra PEPA. In: Priami, C., Ingólfsdóttir, A., Mishra, B., Riis Nielson, H. (eds.) Transactions on Computational Systems Biology VII. LNCS (LNBI), vol. 4230, pp. 1–23. Springer, Heidelberg (2006)
20. Gillespie, D.T.: A general method for numerically simulating the stochastic time evolution of coupled chemical reactions. J. Comp. Phys. 22, 403–434 (1976)
21. Lack, S., Sobocinski, P.: Adhesive and quasiadhesive categories. Theoretical Informatics and Applications 39(3), 511–546 (2005)

One Modelling Formalism & Simulator Is Not Enough! A Perspective for Computational Biology Based on JAMES II

Adelinde M. Uhrmacher, Jan Himmelspach, Matthias Jeschke, Mathias John,
Stefan Leye, Carsten Maus, Mathias Röhl, and Roland Ewald

University of Rostock
18059 Rostock, Germany
Albert-Einstein-Str. 21, D-18059 Rostock, Germany
lin@informatik.uni-rostock.de

Abstract. Diverse modelling formalisms are applied in Computational Biology. Some describe the biological system in a continuous manner, others focus on discrete-event systems, or on a combination of continuous and discrete descriptions. Similarly, there are many simulators that support different formalisms and execution types (e.g. sequential, parallel-distributed) of one and the same model. The latter is often done to increase efficiency, sometimes at the cost of accuracy and level of detail. JAMES II has been developed to support different modelling formalisms and different simulators and their combinations. It is based on a plug-in concept which enables developers to integrate spatial and non-spatial modelling formalisms (e.g. STOCHASTIC π CALCULUS, BETA BINDERS, DEVS, SPACE-π), simulation algorithms (e.g. variants of Gillespie's algorithms (including Tau Leaping and NEXT SUBVOLUME METHOD), SPACE-π simulator, parallel BETA BINDERS simulator) and supporting technologies (e.g. partitioning algorithms, data collection mechanisms, data structures, random number generators) into an existing framework. This eases method development and result evaluation in applied modelling and simulation as well as in modelling and simulation research.

1 Introduction

A Model (M) for a system (S) and an experiment (E) is anything to which E can be applied in order to answer questions about S. This definition that has been coined by Minsky in 1965 [Min65] implies the co-existence of several models for any system. Each model and its design is justified by its specific objectives. Simulation on the other hand can be interpreted as "an experiment performed at a model", as stated by Korn and Wait [KW78]. The term simulation is sometimes used for one simulation run, but more often it refers to the entire experimental setting including many simulation runs and the usage of additional methods for optimization, parameter estimation, sensitivity analysis etc. Each run requires a multitude of steps: e.g., model selection, initialization, defining the observers, selecting the simulation engine, and storing results – to name only a few. Given the long tradition of modelling and simulation, its many facets and the

J. Fisher (Ed.): FMSB 2008, LNBI 5054, pp. 123–138, 2008.

diversity of application areas, it is not surprising that a plethora of different modelling and simulation methods have been developed, and scarcely less simulation tools.

While modelling and simulation has been applied to gain a better understanding of biological systems for well over four decades, broad interest in applying modelling and simulation in cell biology has been renewed by recent developments in experimental methods, e.g. high content screening and microscopy, and has spawned the development of new modelling and simulation methods. Since the millennium, the significance of stochasticity in cellular information processing has become widely accepted, so that stochastic discrete-event simulation has emerged as an established method to complement conventional ordinary differential equations in biochemical simulations. This is also reflected in simulation tools for systems biology that have started to offer at least one discrete-event simulator in addition to numerical integration algorithms, e.g. [ROB05, TKHT04]. The trend to offer more flexibility is not exclusive to the simulation layer. Different parameter estimation methods [HS05] and different possibilities to describe models (e.g., by rules or with BETA BINDERS [GHP07]), are receiving more and more attention as well. Thus, the insight is taking hold in the computational biology realm that a silver bullet does not exist – there are only horses for courses. This diversification and the implied need for a flexible simulation framework is likely to increase over the next years, particularly as, in addition to noise, space is entering the stage of computational biology. In vivo experiments revealed that many intra-cellular effects depend on space, e.g. protein localization, cellular compartments, and molecular crowding [Kho06]. Approaches that support both stochasticity and space are therefore particularly promising [TNAT05, BR06].

The motivation for developing JAMES II (JAva-based Multipurpose Environment for Simulation) has been to support diverse application areas and to facilitate the development of new modelling and simulation methods. JAMES II has been created based on a "Plug'n simulate" [HU07b] concept which enables developers to integrate their ideas into an existing framework and thus eases the development and the evaluation of methods. In the following we will describe basic concepts and some of the current developments to support cell-biological applications in JAMES II.

2   JAMES II– Plug'n Simulate

The simulation framework JAMES II is a lean system consisting of a set of core classes. The core of JAMES II is the central and most rarely changed part of the framework. The main parts are: User interface, Data, Model, Simulator, Simulation, Experiment, and Registry. We used common software engineering techniques for the creation of the framework, e.g. the model-view-controller paradigm [GHJV95] for decoupling the parts, and the abstract factory and factory patterns [GHJV95] for realizing the "Plug'n simulate" approach. Another important design decision was to split model and simulation code completely. Thus, a simulator can access the interface of a model class but a model class is never allowed to access something in a simulator class. This makes it possible to switch the simulation engine (even during runtime) and to exchange the data structures used for the executable models – an essential feature for a flexible framework. In combination with an XML-based model component plugin, this flexibility enables

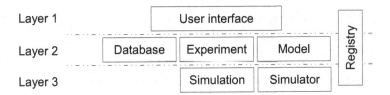

Fig. 1. Packages of the simulation framework JAMES II [HU07b]

the freedom of choice in regards to model data type, simulator code (algorithm as such or parts of the simulation algorithms, e.g. event queues), visualization, and runtime environment. The architecture is sketched in Figure 1. The layers depict the distance of a user from the packages.

Functionality not included in the core classes, especially modelling formalisms and simulation algorithms, can be extended by using plugins. Due to the strict separation between models and simulators, simulation algorithms can be easily exchanged and thus evaluated. This makes the PlugIn mechanism a base for a reliable evaluation of new simulation algorithms. For the integration of a new formalism, one has to create model classes which can represent an instantiated and executable model defined in a certain formalism. Conducting experiments requires at least one additional plugin that provides a simulator. Having created the formalism classes, one can directly start to code models and experiment with them. A prototypical example can be found in [HU07b], where cellular automata are added to the framework. Plugins exist for a number of formalisms, among them variants of DEVS [ZPK00], e.g. ML-DEVS [UEJ+07], and variants of the π CALCULUS, e.g. BETA BINDERS [PQ05] and SPACE-π [JEU08]. If models shall be described in a declarative manner, a model reader can be used, which converts arbitrary model definitions (e.g., from XML files or databases) into executable models [RU06] based on consistency checking of interface descriptions [RM07]. Interface definitions are according to the Unified Modelling Language 2.0 [OMG05]. Thereby, the provisions and requirements of each model component can be explicitly specified, internal details of a component, i.e. the implementation of model behaviour, can be hidden, and direct dependencies between models can be eliminated.

Different formalisms may require different simulation algorithms for their execution. Different hardware infrastructures (e.g. clusters, workstations, the Grid) may impose restrictions or options which should be taken into account by an algorithm. For example, symmetric multiprocessor machines provide fast access to shared memory. Even models described in the same formalism might require different simulators for an efficient simulation, depending on model size and other characteristics. Thus, various simulation plugins have been implemented for JAMES II. As they all provide simulators, i.e. algorithms that execute models, they are integrated via the `simulator` extension point. An extension point subsumes all plugins of one type and provides mechanisms to select the right plugin for a given problem (see fig. 2).

Other extension points provide partitioning and load balancing, random number generation and probability distributions, optimization, parameter estimation, data storage and retrieval, and experiment definitions. JAMES II also offers several extension points

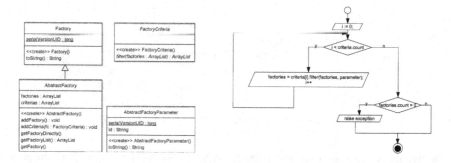

Fig. 2. Factory classes and flow-chart describing the plugin filtering process

for the integration of data structures, among them an extension point for the integration of event queues. Extension points can be defined easily, so their number and the number of implemented plugins grows steadily.

The simulator extension point can be combined with other extension points and forms the basis of a flexible and scalable experimentation layer. Such an experimentation layer can adapt to different types of experiments (e.g. optimization, validation), to different execution schemes (e.g. sequential, parallel distributed), and to different models realized in different modelling formalisms.

The selection process for a simulation algorithm is illustrated in figure 2. Several criteria are subsequently applied to select a suitable simulation algorithm. The first criterion is built-in and selects factories according to the user parametrization of the run. Each plugin is described by a list of factories and a name [HU07b]. For example, selecting a suitable simulator requires to apply a set of specific criteria for factories that can instantiate those for the model at hand. Such a factory also needs to suit the given partition (i.e. whether to compute the model in a distributed or sequential manner), and finally the most efficient factory shall be determined. For this, we strive to take earlier experiences regarding simulator performance into account [EHUar] (see section 4.2).

Experiments with JAMES II show that exchangeable algorithms are the precondition for an efficient execution of experiments [HU07a]. In addition, a subset of these algorithms are reusable across simulation algorithms, even if they have been designed for different formalisms. Thus, the implementation effort required for the creation of algorithms is significantly reduced, see e.g. [HU04]. In addition, the validation and evaluation of simulation algorithms is eased in such a framework: a novel algorithm simply has to be added to the framework and can be compared to all competing solutions, without bias or the need to recode those. The benefits of reusing simulation algorithms or data structures is similar to the arguments that pushes the development of model components. For example, if we define the nucleus as one model component, this allows us to easily replace the model by more abstract, refined, or alternative versions. Thus, a space of interesting hypotheses referring to the system under study can be evaluated without the need to recode the complete model. This freedom of choice with respect to model and simulator configuration is accompanied by an explicit representation of the experimental setting in XML, ensuring repeatability of simulation runs and comparability of the results achieved by alternative models.

3   Modelling Formalisms Supported in JAMES II

Modelling means to structure our knowledge about a given system. Thereby, the modelling formalism plays a crucial role. Whereas continuous modelling formalisms support a macro, population-based view, with its associated continuous, deterministic dynamics progressing at the same speed and being based on real-valued variables, discrete event approaches in contrast tend to focus on a micro perception of systems and their discrete, stochastic dynamics. Variables can be arbitrarily scaled. Sub-systems advance over a continuous time scale, but in steps of variable size. Discrete event models forego assumptions about homogeneous structure or behaviour. For example, DEVS [ZPK00], PETRI NETS [Mur89], STATE CHARTS [Har87] and STOCHASTIC π CALCULUS [Pri95] are formal and generally applicable approaches toward discrete event systems modelling. Their use has been explored in Systems Biology (e.g. [PRSS01, FPH$^+$05, EMRU07]), and depending on their success inspired a broader exploitation and the refinement of methods. The focus in JAMES II has been on discrete event formalisms, with emphasis on DEVS, followed by STOCHASTIC π CALCULUS; lately a first plugin for STATE CHARTS has been implemented.

3.1   Biologically Inspired Variants of DEVS

The strength of DEVS lies in its modular, hierarchical design which provides a basis for developing complex cellular models, whose individual entities are easy to understand if visualized, e.g. by a STATE CHARTS variant. In contrast to STATE CHARTS, the interaction between individual components is explicitly modelled and the interfaces between systems are clearly described by input and output ports. DEVS supports a parallel composition of models via coupling. The general structure of DEVS models is static, a problem that it shares with STATE CHARTS and PETRI NETS. The generation of new reactions, new interactions, or new components is therefore not supported by default. Starting already in the 1980s, several extensions have addressed the problem of variable structures, among them recent developments like DYNDEVS, ρ-DEVS, and ML-DEVS, which are realized as plugins in JAMES II.

   DYNDEVS [Uhr01] is a reflective variant of DEVS which supports dynamic behaviour, composition, and interaction patterns. In ρ-DEVS [UHRE06] dynamic ports and multi-couplings are introduced, whose combination allows models to reflect significant state changes to the outside world and to enable or disable certain interactions at the same time. Unlike DEVS, which realizes an extensional definition of couplings between individual ports, multi-couplings in conjunction with variable ports form an elegant mechanism of dynamic coupling. ML-DEVS [UEJ$^+$07] is our most recent addition to the family of DEVS-formalisms. It inherits from ρ-DEVS variable structures, dynamic ports, and multi-couplings. In addition, it supports an explicit description of macro and micro level. Therefore, the coupled model is equipped with a state and a behaviour of its own, such that the macro level does not appear as a separate unit (an executive) of the coupled model. Secondly, information at macro level can be accessed from micro level and vice versa, which is realized by value couplings and variable ports. Downward and upward activation can be modelled by synchronous activation of multiple micro models and invariants at macro level. Current work is dedicated to applying the formalism to RNA secondary structure prediction and associated regulatory mechanisms.

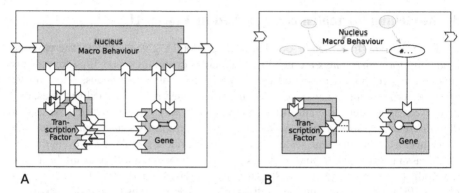

Fig. 3. Comparison of multi-level modelling with DEVS and ML-DEVS. (A) With DEVS the macro level behaviour is modelled at the same hierarchical level as the micro level models. (B) With ML-DEVS the macro dynamics are part of the coupled model. Functions for downward and upward causation reduce the number of explicit couplings needed.

Whereas the biologically inspired extensions of the DEVS formalisms aim to soften the original rigid, modular and hierarchical structure of the formalism, e.g. by introducing variable structures, variable ports, and multi-level interactions, the opposite development can be observed for other formalisms which introduce additional means to structure biological models. These structures are often spatially interpreted, e.g. in terms of cell membranes or compartments. For example, this is the case for BETA BINDERS, which base on STOCHASTIC π CALCULUS.

3.2   Taking Space into Account: SPACE-π

BETA BINDERS [PQ05], as well as MEMBRANE-CALCULI [Car03] or BIOAMBIENTS [RPS+04], address the need to structure space into compartments and confine processes and their interactions spatially. Here, the focus is on indirect, relative space. Given the experimental setups, e.g. confocal microscopy, biologists are particularly interested in a combination of the individual-based approach with absolute space. This has been the motivation for developing SPACE-π [JEU08]. SPACE-π extends the π CALCULUS in a way such that processes are associated with coordinates and individual movement functions. The coordinates are related to a given vector space that provides a norm for calculating distances. A minimum distance is assigned to each channel. As usual, processes can communicate over a channel, but as an additional requirement their distance must be less than or equal to the minimum distance of the channel. Processes move continuously between interactions, as determined by their movement functions. Therefore, SPACE-π is considered to be a hybrid modelling formalism.

Introducing space to the π CALCULUS seems to be straightforward, as it just requires some further restrictions to the standard communication rule. However, an explicit notion of time is needed to describe movement functions. Therefore, an essential component of SPACE-π is a timed version of the π CALCULUS. Although many timed process algebras already exist, the common concept of extending process algebras with time as presented in [NS92] cannot be used. This is because on one hand SPACE-π requires communication to occur as soon as possible, e.g. to describe colliding molecules. On the other hand, the time of an interaction strongly depends on process motion. Since processes can change their motion by communication, one interaction may have an impact on all future events. Therefore, the essential assumption of timed process algebras to consider processes one by one and let them idle for a certain amount of time does not hold in case of SPACE-π. Instead, to assure a valid execution, the global minimum time for the next reaction needs to be calculated after every communication. This makes the approach rather unique in the context of timed process algebras.

The advantage of the SPACE-π approach is that intracellular structures and spatial effects can be represented in a very detailed manner. It is possible to build the trails of membranes or microtubules, to introduce compartments and active transportation processes. Even the impact of molecule sizes and shapes can be modelled. This makes the approach applicable to scenarios that are hard to model with implicit representations of space.

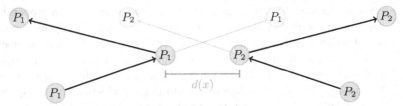

Fig. 4. Communication of two processes: P_1 and P_2 move along some vectors given by their movement functions. If one process is sending and the other receiving on channel x, they can communicate because at some point in time their distance is less than the channel distance $d(x)$. Thereby, they change their movement functions such that they move along different vectors after communicating. Although depicted as simple vectors here, SPACE-π's movement functions can also be used to describe more complex continuous motions.

3.3 Summary

Several plugins exist for various formalisms, including ML-DEVS, SPACE-π, STOCHASTIC π CALCULUS, BETA BINDERS, and chemical reaction networks (described by reaction rules). Although the supported formalisms are quite different, their realization has been facilitated by the core classes in JAMES II and also a reuse of data structures across different formalisms has reduced the implementation effort (see section 5 for further details). However, the architecture and the "Plug'n simulate" concept of JAMES II are even more significant for developing simulation algorithms. For each modelling formalism there is at least one simulator plugin, but many are supported by several simulation algorithms. Up to now, no "multi formalism" approaches are integrated into JAMES II, although such an integration would not pose any problems for the architecture. Some work in this direction is under way, e.g. to combine π CALCULUS and DEVS [MJU07]. These approaches will be strictly separated from the existing realizations (besides the aforementioned reuse of sub-algorithms and data structures) to ensure an efficient execution.

4 Simulation Engines for Computational Biology in JAMES II

The simulation layer of JAMES II focuses on the discrete event world, although some plugins for numerical integration and hybrid simulation exist as well. Gillespie's algorithms [Gil77] and their variants are of particular interest for the area of systems biology and their implementation in JAMES II reflects the idea of re-usable simulation algorithms and data structures. In the following, these and a similar set of process algebra simulators will serve as examples.

In [Gil77], Gillespie introduced two methods to stochastically simulate reaction networks, the Direct Method and the First Reaction Method. Although both methods produce the same results, their performance may differ strongly. This is caused by the use of different operations. For example, the Direct Method needs to generate only two random numbers per iteration, whereas the First Reaction variant needs $r + 1$ random numbers per iteration, r being the number of reactions. Gibson and Bruck propose some enhancements to Gillespie's original algorithms in [GB00], the Next Reaction Method. The new method reduces the time-consuming re-calculation of reaction propensities. A dependency graph is used to identify the reactions that require a propensity update. Furthermore, Gibson and Bruck's approach linearly interpolates the reaction times of all updated reactions, so that the generation of additional random numbers is avoided. Although all SSA approaches work well for small systems and deliver exact stochastic results, they do not perform well when applied to larger problems. This is especially true for systems with concurrent reactions of differing speed (e.g., gene expression and metabolic reactions): If populations of the metabolites are sufficiently high, many iterations (and propensity updates) are needed without any significant changes in any reaction propensity. This problem is equivalent to the challenge of numerically integrating stiff ODE systems. To overcome this problem in the discrete-event realm, a technique called Tau Leaping has been introduced by Gillespie et al. [Gil01]. Tau Leaping approximates the execution of Gillespie's exact approach by leaping forward a time step τ, in which the propensities of all reactions are *approximately* constant. How often each reaction has occurred during this leap can be determined by a Poisson distribution. All reaction occurrences are then executed at once, their propensities are updated, and the algorithm continues by determining the size of the τ leap for the next iteration. One has to solve several problems to implement a suitable Tau Leaping method, as outlined in [TB04, CGP06]. The Direct Method, two implementation variants of the Next Reaction Method, and Tau Leaping have been realized as plugins in JAMES II. They operate on the same model interface and re-use auxiliary data structures like event queues, etc.

4.1 Taking Space into Account

Different approaches toward spatial simulation exist [TNAT05]. At the microscopic level, the fields of molecular dynamics [vGB90] and Brownian dynamics permit the accurate simulation of single interacting particles in continuous space. However, their high computational effort hampers their applications to large scale models containing many thousand particles. Rather than considering individual particles and instead focusing on concentrations of species, partial differential equations operate on a macroscopic level with continuous space and time. As the approach is deterministic, stochastic effects cannot be taken into account easily. The basic algorithm for simulating reactions

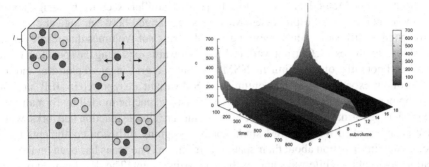

Fig. 5. Left: Discretization of a volume into smaller sub-volumes with side length l. The diffusion of molecules between neighbouring sub-volumes is modelled as a Markov process with $d_i \Delta t = \frac{D_i}{l^2} \cdot \Delta t$ as the probability that a particular molecule of species i performs a diffusion during the infinitesimal small time step Δt (with D_i as the diffusion constant for species i). **Right**: The concentration for sub-volumes of a simple model is plotted for different time stamps. The model consists of a row of 20 sub-volumes with an initial distribution of 1000 molecules of species A in sub-volume 0 and 1000 molecules of species B in sub-volume 19 and a consuming reaction $A + B \rightarrow C$. Molecules A and B diffuse towards the centre where they react and the concentration of molecule C increases.

between chemical species on a mesoscopic level (no individuals, but discrete amount of species elements) is the already mentioned stochastic simulation algorithm (SSA) by Gillespie [Gil77]. One key assumption is that the distribution of the species inside the volume is homogeneous. To simulate systems that do not adhere to this assumption, other approaches that allow to consider compartments and the diffusion of species are necessary, e.g. [Kho06].

A common way of introducing diffusion on mesoscopic level is the partition of space into sub-volumes and the extension of the master equation with a diffusion term, resulting in the reaction-diffusion master equation (RDME) [Gar96]. The solution of the RDME is intractable for all but very simple systems, leading to the development of the Next-Subvolume Method (NSM) [EE04], an algorithm that generates trajectories of an underlying RDME, similarly to SSA sampling the chemical master equation. The Next-Subvolume Method is a discrete-event algorithm for simulating both reactions and diffusion of species within a volume in which particles are distributed inhomogeneously. The volume is partitioned into cubical sub-volumes, each representing a well-stirred system. Events generated by the algorithm are either reactions inside or diffusions between these sub-volumes. A plugin for NSM has been realized in JAMES II. NSM uses a variant of Gillespie's Direct Method to calculate the first event times for all sub-volumes during initialization. Within the main loop, the sub-volume assigned to the smallest next event time is selected and the current event type according to the diffusion and reaction rates is determined. Finally, the event is executed and an update of the model state occurs. Note that the state update is performed only in a small region of the model volume, because either one sub-volume (in the case of a reaction) or two sub-volumes (in the case of a diffusion) are affected and their propensities and next event times have to be updated. With its discretization of space into sub-volumes, the

Next-Subvolume Method lends itself to a distributed parallel execution by assigning sets of sub-volumes to logical processes. Messages exchanged between logical processes correspond to diffusion events involving neighbouring sub-volumes that reside on different logical processes. Current work is directed towards exploring the specific requirements and potential of an optimistic NSM version on a grid-inspired platform and first results of our experiments with an optimistic NSM variant are reported in [JEP+ar]. Our developments based on NSM have not been driven by specific modelling formalisms – the modelling formalisms for the Gillespie variants are simple reaction networks, which have been enriched for the NSM simulator by assigning diffusion coefficients to the species and information about their spatial distribution. The focus has been on the simulation layer taking Gillespie's approach as a starting point. The development of other simulators has been motivated by supporting specific modelling formalisms.

As STOCHASTIC π CALCULUS, SPACE-π, and BETA BINDERS have a common root, basic plugins to support the π CALCULUS have been added to JAMES II. They form a common base for all simulators processing variants of the π CALCULUS. The SPACE-π formalism locates and moves individual processes in absolute space, hence its simulator works in an individual-based manner. With movement functions that are defined to be continuous and could, in principle, even be defined by differential equations, the formalism is clearly hybrid. The hybrid state comprises, besides the processes themselves, all processor positions which change continuously over time. Obviously, this situation provides plenty of challenges for an efficient execution of larger SPACE-π systems. Hence, we developed and implemented an algorithm that approximates movement functions by a sequence of vectors, each of them interpolating the function for δ_t time, resulting in a discretization of the continuous movement of the processes into a sequence of uniform motions. As δ_t can be chosen arbitrarily, the approximation's fidelity can be adjusted seamlessly [JEU08]. The advantage of the simulation is its decent computational complexity, however, if more complex movement functions are involved or a very precise result is necessary, the current solution will not be sufficient and other algorithms will be needed, e.g. approximation methods that adapt their interval size to the slope of the movement functions at hand.

The modelling formalism BETA BINDERS (developed by [PQ05]) provides, similarly to DEVS, means to structure space relatively, e.g. to describe compartments or to distinguish between inter and intra-cellular behaviour by introducing boxes (so called BIOPROCESSES) and explicit interaction sites. A sequential simulator [HLP+06] and a parallel one have been developed for the BETA BINDERS formalism as JAMES II plugins [LPU07]. The stochastic interaction between BIOPROCESSES does not allow an easy definition of lookaheads, which moves the attention to optimistic parallel approaches. However, due to the large effort required in handling the model structure an unbounded optimistic simulation does not appear very promising either. Therefore, we decided to combine conservative and optimistic features in a parallel simulator. Whereas the intra events (that are taking place within one box) are processed in an optimistic manner, the inter events form a kind of barrier and as such are only processed if they are safe. The later is guaranteed by letting all BIOPROCESSES advance up to this barrier [LPU07]. As with all parallel, distributed approaches the question of how to best partition such a model arises, and is subject of ongoing work. JAMES II allows

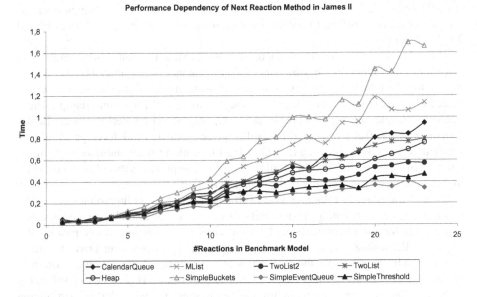

Fig. 6. Performance dependency of the Next Reaction Method regarding the event queue implementation (see [HU07a] for other examples and a discussion of the algorithms). Each setup was replicated 3 - 15 times.

to develop new simulators with very little effort. This is particularly true for simulator families, e.g. simulators for variants of modelling formalisms, e.g. DEVS [HU04] or π CALCULUS. However, also simulators that are quite different share a lot of details. Their re-use reduces programming effort, errors, and, at the same time, increases the chance of an unbiased evaluation. As these details have typically also a significant impact on the efficiency of the simulation engine, they deserve a closer look.

4.2   Details Matter

Simulators are usually not built as monolithic algorithms, but consist of several sub-structures and sub-algorithms. These may have a strong impact on simulation speed and even the accuracy of the produced results (e.g. when considering approximation algorithms). Simulation algorithms may rely on event queues [HU07a], random number generators, or data structures to manage the simulator's state. Parallel and distributed simulators may also require partitioning or load balancing algorithms.

JAMES II provides several plugins for most of those sub-algorithms, including random number generators, probability distributions, event queues, and partitioning algorithms. As an illustration, consider the average runtime of the Next Reaction Method implementation of JAMES II (see section 4) when using different event queue implementations (figure 6).

The benchmark model was a simple 10-species reaction network with arbitrarily many reactions of form $X_i + X_{i+1} \rightarrow X_{i+2}$, but the performance difference is still considerable ($\approx 400\%$ when choosing SimpleEventQueue instead of SimpleBuckets) – even considering different implementations of the *same* algorithm (cf. TwoList and TwoList2 in fig. 6).

Thus, data structures like event queues are of essential importance when evaluating discrete-event simulators. They influence the performance of the simulation engine and may in turn be influenced by certain properties of the model. For example, the performance of an event queue may depend on the distribution of event time stamps. These interdependencies are a general problem in assessing algorithm performance. The design of JAMES II facilitates algorithm performance evaluation and development of new simulation methods by allowing for isolated testing within a fixed, unbiased experimentation environment. The "Plug'n simulate" concept complements this idea by providing means to add new functionality to the system [HU07b].

Partitioning is another domain where sub-algorithms play an important role. Model partitioning, i.e. distributing model entities over the set of available processors, is a prerequisite for any distributed simulation. It aims at minimizing communication between parallel simulation processes and assigning each processor a fair share of entities. Doing it badly may hamper the performance of a distributed simulation algorithm to the point of slower-than-sequential simulation speed. JAMES II provides a partitioning layer [EHU06] to experiment with different partitioning strategies. Several partitioning algorithms have been integrated as plugins (including a wrapper for the METIS [KK98] package) and have been evaluated. Again, experiments showed that the performance gain from using a particular partitioning strategy depends on the simulation algorithm, the model, and the available infrastructure.

Simulation experiments in computational biology are typically not restricted to single simulation runs. They may require thousands of replications and could even include parameter estimation or optimization methods on top (for which again a multitude of methods exist), so that performance issues become even more pressing. Additionally, new algorithms are frequently proposed and need to be evaluated. For example, Cao et al. introduced an optimized version of Gillespie's Direct Method, which is faster than the Next Reaction Method on many problem instances [CLP04]. JAMES II allows to re-validate these findings with custom event queue implementations, since these may have a huge impact on the overall performance [CLP04]. Simulation algorithms that have been built upon the SSA approach, such as the Next Subvolume Method (see section 4), will also benefit from such re-validations, because performance analysis could direct implementation efforts to the most promising code optimizations. These challenges motivated the creation of an infrastructure for simulation algorithm performance analysis in JAMES II [EHUar], which should lead to effective algorithm selection mechanisms in the future.

5   Benefits of JAMES II

The integration of so many aspects into a single modelling and simulation framework requires considerable additional software engineering efforts. From our experiences, these efforts are vastly compensated by time savings from reuse and other benefits, such as unbiased algorithm comparison and validation.

JAMES II provides reuse on different levels. If an additional modelling paradigm shall be supported, only the modelling formalism and at least one simulation algorithm have to be added. Plugins from other extension points, e.g. the experimentation layer with algorithms for optimization, random number generation, partitioning, data sinks and so on, can be reused without any further effort. The actual development of new plugins is supported by predefined algorithms and data structures. For example, simulation algorithms that rely on event queues simply reuse the validated and evaluated ones from the corresponding JAMES II extension point. This reduces the effort for developing new modelling formalisms and simulators significantly.

The integration of new simulation algorithms for existing formalisms is eased as well. Often only small variations between simulators that execute a formalism – e.g. sequentially, in parallel, paced, or unpaced – do exist. To generate a paced variant from an unpaced one usually requires merely a few lines of code, e.g. the Template pattern helped us to develop a set of simulators for DEVS with very little effort [HU04]. Newly developed algorithms can be evaluated and validated easily by comparing their results with those of existing algorithms. Does the new algorithm produce the same results? For which type of model and available infrastructure works the new algorithm best? Developing simulators for new variants of already supported modelling formalisms, e.g. ML-DEVS or SPACE-π, also requires less effort, because parts of existing simulators can still be reused. This is also helpful to identify essential parts and re-occurring patterns in simulation algorithms, which facilitates the conception of new simulation engines.

New data structures and algorithms become available in all situations where their corresponding extension points are used. If a new random number generator or a new event queue is available, it can be directly used in all relevant settings, e.g. when experimenting with stochastic and discrete event models. Thus, new developments are instantly propagated to all potential users. The development of particular data structures and algorithms can be done by domain experts, who have the required knowledge and experience. This is very important, as the field of modelling and simulation subsumes a variety of different expertises. These range from in-depth knowledge about mathematics (random number generation, sensitivity analysis, optimization, statistics, graph theory) and computer science (databases, compilers, efficient data structures and algorithms in general) up to application-specific extensions required for certain domains, e.g. the use of metaphors in biology.

6 Conclusion

The purpose of JAMES II is twofold. On the one hand its flexibility and scalability is aimed at supporting a wide range of different application areas and different requirements. This feature has shown to be of particular value in research areas like Computational Biology, which is characterized by the wish to experiment with diverse model and simulation types. On the other hand JAMES II's virtue lies in facilitating the development and evaluation of new modelling and simulation methods.

Thus, to support application studies and methodological studies equally is the strength of JAMES II, which is also reflected in current projects. Some of them are directed towards applications, e.g. models of the Wnt signalling pathway, or gene regulatory mechanisms. Others are concerned with the development of new modelling and simulation methods, including modelling formalisms like ML-DEVS and SPACE-π, or new simulators like the parallel optimistic versions for BETA BINDERS and NEXT SUBVOLUME METHOD, and the combination of modelling formalisms and simulator engines. Both endeavours are tightly connected, as new requirements drive the development of new methods. Both types of our experiments are directed towards a better understanding of complex systems, i.e. biological systems and simulation systems, and their many interdependencies.

Acknowledgements

This research is supported by the DFG (German Research Foundation).

References

[BR06]   Broderick, G., Rubin, E.: The realistic modeling of biological systems: A workshop synopsis. ComPlexUs Modeling in Systems Biology, Social Cognitive and Information Science 3(4), 217–230 (2006)

[Car03]   Cardelli, L.: Membrane interactions. In: BioConcur 2003, Workshop on Concurrent Models in Molecular Biology (2003)

[CGP06]   Cao, Y., Gillespie, D.T., Petzold, L.R.: Efficient step size selection for the tau-leaping simulation method. J. Chem. Phys. 124, 044109 (2006)

[CLP04]   Cao, Y., Li, H., Petzold, L.: Efficient formulation of the stochastic simulation algorithm for chemically reacting systems. The Journal of Chemical Physics 121(9), 4059–4067 (2004)

[EE04]   Elf, J., Ehrenberg, M.: Spontaneous separation of bi-stable biochemical systems into spatial domains of opposite phases. Syst. Biol (Stevenage) 1(2), 230–236 (2004)

[EHU06]   Ewald, R., Himmelspach, J., Uhrmacher, A.M.: Embedding a non-fragmenting partitioning algorithm for hierarchical models into the partitioning layer of James II. In: WSC 2006: Proceedings of the 38th conference on Winter simulation (2006)

[EHUar]   Ewald, R., Himmelspach, J., Uhrmacher, A.M.: An algorithm selection approach for simulation systems. In: Proceedings of the 22nd ACM/IEEE/SCS Workshop on Principles of Advanced and Distributed Simulation (PADS 2008) (to appear, 2008)

[EMRU07]   Ewald, R., Maus, C., Rolfs, A., Uhrmacher, A.M.: Discrete event modelling and simulation in systems biology. Journal of Simulation 1(2), 81–96 (2007)

[FPH+05]   Fisher, J., Piterman, N., Hubbard, J., Stern, M., Harel, D.: Computational insights into C. elegans vulval development. PNAS 102(5), 1951–1956 (2005)

[Gar96]   Gardiner, C.W.: Handbook of Stochastic Methods: For Physics, Chemistry and the Natural Sciences (Springer Series in Synergetics). Springer, Heidelberg (1996)

[GB00]   Gibson, M.A., Bruck, J.: Efficient Exact Stochastic Simulation of Chemical Systems with Many Species and Many Channels. J. Chem. Physics 104, 1876–1889 (2000)

[GHJV95]   Gamma, E., Helm, R., Johnson, R., Vlissides, J.: Design Patterns: elements of reusable object-oriented software. Addison-Wesley, Reading (1995)

[GHP07]   Guerriero, M.L., Heath, J.K., Priami, C.: An automated translation from a narrative language for biological modelling into process algebra. In: Calder, M., Gilmore, S. (eds.) CMSB 2007. LNCS (LNBI), vol. 4695, pp. 136–151. Springer, Heidelberg (2007)

[Gil77]   Gillespie, D.T.: Exact Stochastic Simulation of Coupled Chemical Reactions. The Journal of Physical Chemistry B 81(25), 2340–2361 (1977)

[Gil01]   Gillespie, D.T.: Approximate accelerated stochastic simulation of chemically reacting systems. The Journal of Chemical Physics (2001)

[Har87]   Harel, D.: Statecharts: A Visual Formalism for Complex Systems. Science of Computer Programming 8(3), 231–274 (1987)

[HLP⁺06]   Himmelspach, J., Lecca, P., Prandi, D., Priami, C., Quaglia, P., Uhrmacher, A.M.: Developing an hierarchical simulator for beta-binders. In: 20th Workshop on Principles of Advanced and Distributed Simulation (PADS 2006), pp. 92–102. IEEE Computer Society, Los Alamitos (2006)

[HS05]   Jirstrand, M., Schmidt, H.: Systems biology toolbox for matlab: A computational platform for research in systems biology. Bioinformatics (2005)

[HU04]   Himmelspach, J., Uhrmacher, A.M.: A component-based simulation layer for james. In: ACM Press (ed.): PADS 2004: Proceedings of the eighteenth workshop on Parallel and distributed simulation, pp. 115–122. IEEE Computer Society, Los Alamitos (2004)

[HU07a]   Himmelspach, J., Uhrmacher, A.M.: The event queue problem and pdevs. In: Proceedings of the SpringSim 2007, DEVS Integrative M&S Symposium, pp. 257–264. SCS (2007)

[HU07b]   Himmelspach, J., Uhrmacher, A.M.: Plug'n simulate. In: Proceedings of the Spring Simulation Multiconference, pp. 137–143. IEEE Computer Society, Los Alamitos (2007)

[JEP⁺ar]   Jeschke, M., Ewald, R., Park, A., Fujimoto, R., Uhrmacher, A.M.: Parallel and distributed spatial simulation of chemical reactions. In: Proceedings of the 22nd ACM/IEEE/SCS Workshop on Principles of Advanced and Distributed Simulation (PADS 2008) (to appear, 2008)

[JEU08]   John, M., Ewald, R., Uhrmacher, A.M.: A spatial extension to the pi calculus. In: Proc. of the 1st Workshop From Biology To Concurrency and back (FBTC 2007). Electronic Notes in Theoretical Computer Science, vol. 194, pp. 133–148 (2008)

[Kho06]   Kholodenko, B.N.: Cell-signalling dynamics in time and space. Nature Reviews Molecular Cell Biology 7(3), 165–176 (2006)

[KK98]   Karypis, G., Kumar, V.: MeTis: A Software Package for Partitioning Unstructured Graphs, Partitioning Meshes, and Computing Fill-Reducing Orderings of Sparse Matrices (Version 4.0) (September 1998)

[KW78]   Korn, G.A., Wait, J.V.: Digital continuous-system simulation. Prentice-Hall, Englewood Cliffs (1978)

[LPU07]   Leye, S., Priami, C., Uhrmacher, A.M.: A parallel beta-binders simulator. Technical Report 17/2007, The Microsoft Research - University of Trento Centre for Computational and Systems Biology (2007)

[Min65]   Minsky, M.: Models, minds, machines. In: Proc. IFIP Congress, pp. 45–49 (1965)

[MJU07]   Maus, C., John, M., Uhrmacher, A.M.: A multi-level and multi-formalism approach for model composition in systems biology. In: Conference on Computational Methods in Systems Biology, Edinburgh, Poster (2007)

[Mur89]   Murata, T.: Petri Nets: Properties, Analysis and Applications. Proceedings of the IEEE 77(4), 541–574 (1989)

[NS92]   Nicollin, X., Sifakis, J.: An Overview and Synthesis on Timed Process Alge-
             bras. In: Larsen, K.G., Skou, A. (eds.) CAV 1991. LNCS, vol. 575, pp. 376–398.
             Springer, Heidelberg (1992)
[OMG05]   OMG. UML superstructure specification version 2.0 (document formal/05-07-04)
             (July 2005),
             http://www.omg.org/cgi-bin/doc?formal/05-07-04
[PQ05]   Priami, C., Quaglia, P.: Beta binders for biological interactions. In: Danos, V.,
             Schachter, V. (eds.) CMSB 2004. LNCS (LNBI), vol. 3082, pp. 20–33 Springer,
             Heidelberg (2005)
[Pri95]   Priami, C.: Stochastic π-calculus. The Computer Journal 38(6), 578–589 (1995)
[PRSS01]   Priami, C., Regev, A., Shapiro, E., Silvermann, W.: Application of a stochastic
             name-passing calculus to representation and simulation of molecular processes.
             Information Processing Letters 80, 25–31 (2001)
[RM07]   Röhl, M., Morgenstern, S.: Composing simulation models using interface defini-
             tions based on web service descriptions. In: WSC 2007, pp. 815–822 (2007)
[ROB05]   Ramsey, S., Orrell, D., Bolouri, H.: Dizzy: Stochastic simulation of large scale
             genetic regulatory networks. Journal of Bioinformatics and Computational Biol-
             ogy 01(13), 415–436 (2005)
[RPS+04]   Regev, A., Panina, E.M., Silverman, W., Cardelli, L., Shapiro, E.: BioAmbients:
             an abstraction for biological compartments. Theor. Comput. Sci. 325(1), 141–167
             (2004)
[RU06]   Röhl, M., Uhrmacher, A.M.: Composing simulations from xml-specified model
             components. In: Proceedings of the Winter Simulation Conference 2006, pp.
             1083–1090. ACM, New York (2006)
[TB04]   Tian, T., Burrage, K.: Binomial leap methods for simulating stochastic chemical
             kinetics. The Journal of Chemical Physics 121(10356), 10356–10364 (2004)
[TKHT04]   Takahashi, K., Kaizu, K., Hu, B., Tomita, M.: A multi-algorithm, multi-timescale
             method for cell simulation. Bioinformatics 20, 538–546 (2004)
[TNAT05]   Takahashi, K., Nanda, S., Arjunan, V., Tomita, M.: Space in systems biology of
             signaling pathways: towards intracellular molecular crowding in silico. FEBS let-
             ters 579(8), 1783–1788 (2005)
[UEJ+07]   Uhrmacher, A.M., Ewald, R., John, M., Maus, C., Jeschke, M., Biermann, S.:
             Combining micro and macro-modeling in devs for computational biology. In:
             Proc. of the 2007 Winter Simulation Conference, pp. 871–880 (2007)
[Uhr01]   Uhrmacher, A.M.: Dynamic structures in modeling and simulation - a reflective
             approach. ACM Transactions on Modeling and Simulation 11(2), 206–232 (2001)
[UHRE06]   Uhrmacher, A.M., Himmelspach, J., Röhl, M., Ewald, R.: Introducing variable
             ports and multi-couplings for cell biological modeling in devs. In: Proc. of the
             2006 Winter Simulation Conference, pp. 832–840 (2006)
[vGB90]   van Gunsteren, W.F., Berendsen, H.J.: Computer simulation of molecular dy-
             namics: Methodology, applications, and perspectives in chemistry. Angewandte
             Chemie International Edition in English 29(9), 992–1023 (1990)
[ZPK00]   Zeigler, B.P., Praehofer, H., Kim, T.G.: Theory of Modeling and Simulation. Aca-
             demic Press, London (2000)

Author Index

Lecture Notes in Bioinformatics

Vol. 3692: R. Casadio, G. Myers (Eds.), Algorithms in Bioinformatics. X, 436 pages. 2005.

Vol. 3680: C. Priami, A. Zelikovsky (Eds.), Transactions on Computational Systems Biology II. IX, 153 pages. 2005.

Vol. 3678: A. McLysaght, D.H. Huson (Eds.), Comparative Genomics. VIII, 167 pages. 2005.

Vol. 3615: B. Ludäscher, L. Raschid (Eds.), Data Integration in the Life Sciences. XII, 344 pages. 2005.

Vol. 3594: J.C. Setubal, S. Verjovski-Almeida (Eds.), Advances in Bioinformatics and Computational Biology. XIV, 258 pages. 2005.

Vol. 3500: S. Miyano, J. Mesirov, S. Kasif, S. Istrail, P.A. Pevzner, M. Waterman (Eds.), Research in Computational Molecular Biology. XVII, 632 pages. 2005.

Vol. 3388: J. Lagergren (Ed.), Comparative Genomics. VII, 133 pages. 2005.

Vol. 3380: C. Priami (Ed.), Transactions on Computational Systems Biology I. IX, 111 pages. 2005.

Vol. 3370: A. Konagaya, K. Satou (Eds.), Grid Computing in Life Science. X, 188 pages. 2005.

Vol. 3318: E. Eskin, C. Workman (Eds.), Regulatory Genomics. VII, 115 pages. 2005.

Vol. 3240: I. Jonassen, J. Kim (Eds.), Algorithms in Bioinformatics. IX, 476 pages. 2004.

Vol. 3082: V. Danos, V. Schachter (Eds.), Computational Methods in Systems Biology. IX, 280 pages. 2005.

Vol. 2994: E. Rahm (Ed.), Data Integration in the Life Sciences. X, 221 pages. 2004.

Vol. 2983: S. Istrail, M.S. Waterman, A. Clark (Eds.), Computational Methods for SNPs and Haplotype Inference. IX, 153 pages. 2004.

Vol. 2812: G. Benson, R.D.M. Page (Eds.), Algorithms in Bioinformatics. X, 528 pages. 2003.

Vol. 2666: C. Guerra, S. Istrail (Eds.), Mathematical Methods for Protein Structure Analysis and Design. XI, 157 pages. 2003.